The Real and the Imaginary

The Real and the Imaginary

A New Approach to Physics

Edited by Jean E. Charon

An ICUS Book

Paragon House Publishers
New York

Published in the United States by

PARAGON HOUSE PUBLISHERS
2 Hammarskjöld Plaza
New York, New York 10017

An International Conference on the Unity of the Sciences Book.

Library of Congress Cataloging-in-Publication Data

Colloque international pluridisciplinaire sur le rôle de l'esprit en science (2nd : 1984 : Washington, D.C.)
The Real and The Imaginary.

Translation of: Imaginaire et réalité
"An ICUS book."
Bibliography:
Includes index.
1. Physics—Philosophy—Congresses. 2. Science—Philosophy—Congresses. 3. Imagination—Congresses. 4. Reality—Congresses. I. Charon, Jean E. II. Title.
QC5.56.C65 1984 530'.01 86–25326
ISBN 0–89226–027–0

Contents

Introduction

Jean E. Charon

Some professors and scholars attending the Thirteenth ICUS International Conference, which was as usual devoted to the examination of science and value, might have been somewhat surprised when they discovered that one of the committees had chosen for its topic "Imaginary and Reality." Reality is a truly scientific subject; but what does the imaginary have to do with science? According to the dictionary, the imaginary only exists in the imagination and only refers to objects which are not real. Does science have something to do with "what is not real"? Are physics and biology today becoming preoccupied with "ghosts" or "fantasy"?

Curiously enough, although "objectivity" is still (and indeed by definition) the fundamental method of science, the imaginary has now become a *scientific* field of investigation. The imaginary has now left psychology and philosophy to become a subject of research in the "hard" disciplines, like biology and physics. Obviously, however, a somewhat enlarged meaning has been given to the word "imaginary." It here designates all that affects the thought processes of the observer scientifically looking at the external world and trying to give a more complete representation of it. Briefly speaking, the topic "Imaginary and Reality," as submitted to a "scientific" gathering, recognizes that the universe around us roughly consists of matter on one hand, and thought, observing the behavior of this matter, on the other hand. For the last three centuries only matter had been the center of investigation of physics and biology, but today the representation of matter includes such minute (though such important) details that it is no longer possible to get still "deeper" in the representation without saying something precise about the characteristics of thought (the observer) representing this matter. The more we encounter pure matter's behavior, the more it is necessary to obtain a better knowledge of the "interaction" between thought (the observer) and matter (the observed).

For such a scientific investigation of the mind's processes, science should encompass a larger, *interdisciplinary* meaning. So little is yet known about mind that we have to examine it using many traditionally separate disciplines: to philosophy and psychology, which traditionally have focused around the mind problem; but now to the "hard" sciences as well, disciplines such as biology and physics.

Let us look, for instance, to the concept (now familiar to all biologists) of "transmaterial" data circulating in the brain structure, along the neurons and the synapses: what is exactly meant by this word "transmaterial"? Does it mean that the biologists have nowadays to take into account, in their representation of the brain, entities which can no more be qualified as "material"? But, if this is the case, how can science represent entities that have nothing to do with physical or chemical concepts?

And consider contemporary physicist John Wheeler, director of the Department of Theoretical Physics at the University of Texas at Austin, who recently declared that we live in a "participatory universe," meaning that the behavior of an individual material particle may be directly influenced by the observer's mind considering that particle's behavior and trying to represent it in a mathematical formalism. We have long known that any "observer" could influence the behavior of a *living* object, like an animal or another human individual, for instance: but is this still true when the observer considers the behavior of apparently "inert" objects, like the elementary particles of physics? In that case, where is the right delimitation between living and inert objects? Is that a question which could be answered through a better knowledge of "pure" matter? Are there some properties of mind associated with this "pure" matter?

It is clear that *interdisciplinary* research is needed on such a fundamental topic, and moreover, this research is *urgently* needed: answers to questions concerning mind processes and the relations between mind and matter would not only contribute to a better representation of our universe, but also to a better understanding of *each of us* as individuals since we are first of all "thinking" beings, not mere material bodies. We enter an era when science will direct its investigations toward a *unified* physical and spiritual world.

These are some of the reasons for the interdisciplinary meeting considering the "Imaginary and Reality." And we bet that in coming years, as one of the participants (Gilbert Durand) remarked, these meetings will become more and more "a time of reunion between the Imaginary of Science and the Science of Imaginary."

1.
What Could be a Mind Able to Conceive a Brain Able to Produce a Mind?

Edgar Morin

Who can doubt the presence of the mind? If we abandon the illusion that sees in the soul an "immaterial" substance, that would not be denying its existence, but on the contrary would be beginning to recognize the complexity, richness, and unfathomable depth of its genetic and cultural inheritance, like personal experience, conscious or not, that together make up the beings we are, unique and indisputable witnesses of ourselves.

Jacques Monod
(*Chance and Necessity*)

There is no borderline between life and the mind.

Etienne Wolff

If, in our time, someone decided to deduce the intellectual or spiritual phenomenology from glandular activity, he could be assured a priori *of the esteem and concentration of his audience; if, on the other hand, someone else thought he saw, in the atomic disintegration of celestial matter, an emanation of the creator-spirit of the world, this same audience could only deplore the mental aberration of the author. However, these two explanations are equally logical, equally metaphysical, equally arbitrary, and equally symbolic . . . . The hypothesis of mind is no more fantastic than that of matter.*

C.G. Jung

The extraordinary problem

The brain is without doubt the least well-known of all the continents explored by scientific knowledge. Yet this unknown is the source of all knowledge. And presenting even more uncertainty is the abyss of ignorance concerning the nature of the relationship between the brain and the mind. There is even incredible disagreement about the sense and relevance of the word mind, yet the two notions of brain and mind are as inextricably intertwined as the Gordian knot. Around this knot turn visions of the world, of man, and of knowledge, that can only be cut by a strike of a barbarian sword.

In a sense, mind and brain are two aspects of the same thing. But at the same time, what an ontological, logical, epistemological gap exists between them! These jelly-like brains, what have they to do with ideas, religion, philosophy, goodness, pity, love, poetry, freedom? This limp mass, as astonishing as the termite queen's abdomen, how can it generate discourse, meditation, knowledge? How can it be that this substance that has no feeling can cause us pain? What does this unfeeling magma know of the misery and happiness it lets us feel?

The mind has an incredible natural blindness to this brain, without which it would not exist. What does the mind know about the brain? Nothing spontaneously. It was only the practice of medicine that discovered, with Hippocrates, the spiritual role of the brain, and it was only experimental knowledge that began to explore it.

Thus there is simultaneously an ontological abyss and a mutual mystery between, on the one hand, a cerebral organ composed of regions and strata which bring together tens of billions of neurons, themselves linked by networks of unimaginable complexity, and animated by electrical and chemical processes, and on the other hand the image, the idea, and thought. The mind knows nothing, in itself, of the brain that produces it, and the brain in turn knows nothing of the mind it produces. They know without knowing each other. Their unity produces awareness even though they are unaware of it. Of course one can speak separately about the mind and the brain, and they are studied independently, the mind in social science departments, and the brain in the biological science departments. Psycho-sciences and neuro-sciences do not usually communicate, even though the most basic question for both is their relationship.

The Great Schism

The mind and brain thus remain two notions condemned either to complete disunity or to reductionism (of brain to the mind). The Great Disunity that has reigned over western culture since the seventeenth century has relegated the brain to the domain of science, and made it obey the deterministic and mechanistic laws of matter, while the mind exists insubstantial, creative and free in the domain of philosophy and the humanities. When these two domains meet, they wage the great metaphysical war of the free mind against deterministic Nature, with the main battle being fought on the terrain of the mind-brain relationship.

These two metaphysical obsessions of materialism and spirituality have coincidentally spurred various research methods and infected any reflection on the mind and the brain, which has been rent between the alternatives of materialism or spirituality. These two attitudes must, however, be understood. It is understandable that the mind first appeared to be a superior entity because it is conscious and directs thought, decisions, and actions and that the world seems animated by a spiritual principle. Starting in the nineteenth century, however, Mind was forced to descend from heaven and undergo, in the universe of science, a terrible degradation. The Mind of God, according to the Scripture and to theological reason, created the world, life and men. The creation descended from superior to inferior. But with Lamarck and then with Darwin, a reversal occurred: everything now began from the lowest point and then grew upwards, evolving toward the heights, and the mind became the greatest fruit of evolution, no longer the first principle of creation. At the same time, the science of the nineteenth century spread materialistic determinism into all realms at the expense of the mind.

From then on, it is natural that there can be triumphally affirmed the material monism of a Vogt, for example, who said that the brain "excretes feelings like the kidneys excrete urine." The mind, from this viewpoint, can only be like the Self, the ghost of the machine.

At any rate, the discoveries of the nineteenth and twentieth centuries have revealed the skills and the performance of the brain in organizing memory, language, and thought and thus have caused the ebb of spirituality. In this defeat, however, Bergson fights his battle of the Marne, which creates a stable front of resistance. Bergson recognizes the beneficial results of brain research, but only to confirm his idea that the mind manifests itself in all parts of the brain, which is only an "image" produced by the mind. From this point on, while the materialists continue to call the mind an epiphenomenon, the spiritually oriented people try to conceive of the brain as a sort of network that captures "transmaterial" messages which are transmitted in a psychic or information field. The brain does not "produce" the mind, but "detects" it (Burt, Eccles). The information that penetrates the senses "materializes" itself in chemical substances and neuronal changes, which register physically the symbolic meaning of sensory perceptions.

These spiritual conceptions, forced to compromise with the material reality of the brain, result in a collaborative dualism or interaction, which accepts the fact that spiritual reality requires the cooperation of a material reality to carry out its operations.

This spiritual resistance on the mind-brain front is bolstered by a reversal in the situation where materialism had won its most decisive victory, that is, in the basis of physical reality. The collapse of material substantiality at the subatomic level gave rise to an enigma and a mystery, and there spirituality rushed in, hoping to reconquer the world not in spite of, but due to the progress of science.

Here we cannot opt for either one of the two hypotheses in conflict, but we can support the question asked by each one:

1. The first question has become absolutely crucial since the last century: "the mind, after having explained everything, is itself in need of explanation" (Bateson 1980, 98).

2. The second question is raised by contemporary physics: matter, after having explained everything, must be explained.

The brain cannot explain the mind; furthermore, it needs the mind to explain itself. Thus the brain cannot conceive a brain except via the mind, which itself cannot conceive a mind except via the brain.

From now on, then, we must refuse to consent to the elimination of the mind by the brain (science) and the brain by the mind (philosophy). Each one of these paths leads to a strange aberration. The philosophers' pure mind needs philosophers endowed with brains. On the other hand, it is the mind and consciousness of the scientists that conceived of a universe without mind and without consciousness.

Whence this double blindness? How can there be such willful unseeing in each of these metaphysics? We believe it is because both are suffering the effects of the same paradigm, the paradigm that forces an absolutely clear choice between the one or the double, that is between monism or dualism. These two antagonistic conceptions of materialism or spiritualism in their pretensions for hegemony follow the first paradigm; thus, materialism reduces everything spiritual to a pure emanation of matter, and spiritualism reduces everything material to an emanation of the mind. Whenever these two conceptions lose their absolutist pretensions and agree to compromise, then they are willing to consider two substances, different in nature and in reality, but led to communicate with each other and linked by causal interactions. In this way, an entire modern tradition begins with Descartes, passes through Bergson, and ends with Popper-Eccles.

What is thus very difficult to conceive, due to the lack of a paradigm able to confront this complexity, is the *uniduality* of the brain-mind relationship, that is, their concurrent unity and duality of existence and reality. Any consideration of this duality has a tendency to destroy this profound unity, and any consideration of their profound unity tends to destroy their insurmountable duality (such is the case, as we will see later on, with the "double aspect" theories or theories of co-reference).

Uniduality

The brain demands recognition: the more we learn about it, the more we recognize its powerful organizational reality. The mind, for its part, retreats, is worn away, breaks up, becomes dependent in the least of its operations on molecular interactions,[1] loses any autonomy, and evaporates as a result of a deterministic or reductionist principle.

We must recognize, however, even as an epiphenomenon, the unique reality of the mind, which takes the form of images, ideas, concepts, thought. The things of the mind, even though dependent on the brain, are nevertheless objects of a kind: language, logic, ideas, theory. By virtue of this, indubitably, the human mind really does exist. But what kind of existence does it have? What reality does it enjoy? This can only be understood through an articulation, not an exclusion, of the reality and existence of the brain, and this articulation cannot be abstract and separate, as we will see, from either the whole human being, or from culture and society.

Thus we must confront uniduality, which is the problem of the unity and inseparability of the two differing and irreducible realities. No operation of the mind can escape a local and general activity of the brain, and we must give up any idea of a human psychic phenomenon independent of a bio-physical phenomenon.

At this point, the paradox of uniduality reappears: what is inseparable here are the two universes, two types of organizations. On one hand, via the brain-organ, are the cabling and neuronal networks, the inter-neuronal activities and communications, the entire biological organization, and, underlying it, the physico-chemical processes. On the other hand, via the mind, we have ideas, language, logic, communications and socio-cultural organization, that is to say, transcerebral and metabiological processes. Whence, once more, the ontological and logical gap between the bio-physical (brain) and the sociocultural (mind). What relationship could there possibly be at this level? Does this relationship go from the product to the producer? From effect to cause?

Here again there appears on a first level the indisputable relationship of dependence, both between the mind and brain, and between the socio-cultural and the bio-physical. We can act chemically, electrically, anatomically on all the characteristics of the mind, either to modify them, stimulate them, or to suppress them. We can destroy consciousness by sections and lesions of certain parts of the brain, we can modify states of consciousness through the use of drugs, we can manipulate consciousness and render it unaware of the manipulations it is undergoing; electric or chemical interference to certain zones of the brain can provoke visions, hallucinations, feelings, emotions, all of which demonstrate that the mind can be made to submit blindly to physico-chemical modifications. Furthermore, we are constantly learning that psychological states are strictly dependent on a lack or excess of one or another molecular constituent (thus depression corresponds to a reduction of serotin in the brain).

But here is a product that can react upon its producer, an effect that can react on its cause. All the psychosomatic phenomena point to a causal circularity, where the mind itself acts on the brain, and via the brain on the organism, that is, on the bio-chemical-physical processes themselves.

Even more: the extraordinary psycho-cultural phenomenon that is faith can, via the brain, motivate the healing or ruin of the organism on which the psyche and culture depend, by provoking in magical fashion cure or death (thus placebos, witches, and miracles heal one third of all sick people). The consciousness of a taboo, of a spell, of a curse can kill. We can even note, in the example of yogis able to control by asceticism, or the discipline of the will, muscles and organs like the heart, the problem of the relative autonomy of the mind. And all of these examples—belief, faith, willpower—indicate both the relative autonomy and the specific, undeniable reality of the world of the mind, which is, at the same time, the world of culture: myths, religions, ideas. . . .

Even more again: we recognize that the mind experiences determining influences, which are not only those of the cerebral organization nor those of events lived or experienced by the individual, especially during his childhood, but also those from a cultural imprinting, unchangeable influences taking place in the family, school, university, culture. Thus, via the mind, these socio-cultural determinations are imposed in the very construction of the links and circuits between synapses, creating, in the geography of the brain itself, routes, tracks and pathways traced and marked by culture.

Furthermore, we must understand that knowledge, even of bio-cerebral determinations, comes both from the mind (researchers working in the neuro-sciences) and from socio-cultural reality, which is the scientific institution. And finally, human intelligence also generates interferences or manipulations in the brain. Even more, there are awesome transcerebral entities which constitute modern nation-states, which attempt to control all brains in order to subdue all minds!

Let us not forget that it was the human mind that first went in search of the brain; later, psycho-cultural activity discovered that the brain produced the mind. The mind formulated hypotheses and sought clarification of its own relationship with the brain through scientific effort and through cultural, historical and social effort, as well. It is through the study of the brain that the mind confirms itself in its ability to discover and to think, and then to govern and control its own source. The brain allows itself to be manipulated by its spiritual epiphenomenon. Or rather, it is the uniduality of mind/brain, in the deepest part of a socio-cultural collectivity that can only organize itself around interactions between minds/brains, that operates these controls bearing on the cerebral aspect, which thus can no longer be regarded as the supreme controller of the spiritual and the cultural.

The relationship of brain to mind cannot be thought of simply as that of producer to product, of cause to effect, of the emanating to the emanated. There is a relative autonomy, reciprocal action, mutual control that must be understood and explained. How? By a return to dualism? By the elaboration of a new monism, which would be neither spiritual nor material? Again there intervenes the mutilating paradigmatic demand that we choose between the two terms of the alternative. This dualism is one of the complementarity between two heterogeneous realities, of which one has an absolute need for the support of the other. Thus there is complementarity between the material entities (brain, organism) and the "transmaterial" entities (Delgado) that are information, symbols, and values, and we can imagine a morphological balance between the neuro-cerebral forms and the symbolic-informational forms.[2]

The New Monism

The new monism notes the impossibility of unity, either on the material basis (brain) or the spiritual basis (mind). From this spring the theses of identitism and co-reference. The theory of identitism posits the identity of cerebral and mental states, which are co-referents for the same object. They both refer to one and the same identity.

It is precisely this contradiction that we must neither hide nor avoid, but face squarely. But affirming it would be to betray the paradigm of supposedly logical thought, for which the appearance of a contradiction can only mean the appearance of an error. From the point of view of complex thought, however, the very complexity of reality is manifested in the development of relationships which, from a conceptual point of view, are at once complementary, competitive and antagonistic. It is the same for the idea of *unitas multiplex* and for the idea of uniduality that we proposed. It is the same in the relationship between the mind and the brain, as Andre Bourguignon lucidly observed: "The solution to the matter-mind relationship can only be contradictory: the body (nervous encephalic activity) and the mind (psychic activity) are at the same time identical, equivalent and different, distinct. A solution demands that one never prefer one of the terms of the contradition to the other, especially in scientific research" (Bourguignon 1981, 397).

In considering the contradiction, we have been led to the paradoxal circularity of the two notions of brain and mind. Indeed, if the brain can be considered an instrument of thought, thought can be considered a tool of the brain. The mind can be seen as a typical product of the brain, but even the notion of brain must be conceived of as a typical product of the mind: the brain is a conceptualization engendered by the mind. Sensorial reality, which has reached sensory terminals and been translated, coded, and transmitted to the brain, is itself recognized by our mind after a long process of intellectual elaboration, the fruit of the development of scientific knowledge in the historical conditions which permitted this development. Thus, in order to understand that our senses are affected by variations in energy (mechanical modifications for hearing and touch, chemical modifications for the sense of smell, photonic modifications for our vision), a previous elaboration of the concept of energy was first necessary; this invisible and abstract entity only becomes aware of itself through its modifications. This means that, even to conceive of a physical reality which affects our brain and allows our brain to know the world around it, it was necessary at the outset to develop a spiritual fabric saturated with cultural substance. A specific work of the mind was needed at the very heart of a culture to allow it to conceive of the obvious physical reality that affects our senses and the obvious biological reality that constitutes our brain. Thus, as objective as they may be, the physical universe, its modifications of energy, our senses, and our brains are themselves but representations of our mind. This unidual activity of brain/mind produces a conception of physical reality at the same time it recognizes its own reality.

Thus it is not only the mind that directs us to the brain and to the physical reality that gives us knowledge; it is also physical reality and the brain that direct us to the reality of the mind, which is necessary in order to conceive of them both. In a sense, the brain and the physical world are both products of the mind, provided we do not forget that, from the point of view of the evolution of the physical world, the biological world and especially the animal world, the mind is the product of the brain.

What appears, then, is a type of circle, where, each term hellishly directs us to the other term, and then is metamorphosed into the other. Holton cites the excellent example of Hermann Weyl: "as we know, the chalk of the blackboard is composed of molecules which are made up of elementary particles. In analyzing what theoretical physics means by such words, we have seen that these physical objects melt into a symbolism which can be manipulated by certain rules; but these symbols, in turn, are finally only substantial signs written in chalk on a blackboard."

It is indeed a metaphorical circle between the brain and the mind. The mind is an outgrowth of the brain, which is itself a representation of the mind. The mind is, as an activity, a product of the brain, itself a biological organ that can only be recognized by the mind in and through a cultural organization. Thus, in a circle always mediated by a society and a culture, the brain can only conceive of itself via the mind, but the mind can conceive of itself only via the brain. Whence the paradox we must finally face: what is a mind that can conceive of the brain that produces it, and what is the brain that can produce a mind that conceives of it?

It is now clear to us that any concept of the brain that does not take into account the paradoxical Gordian knot of the brain-mind relationship would be distorting. The spiritualist vision, if it ignores the brain, can not do justice to the mind itself. But any knowledge of the brain that cannot acknowledge the mind, or worse yet, that excludes it, becomes absurd because it becomes incapable of acknowledging iself. Scientific knowledge can no longer deny the mind without which it would not exist. But above all, and ironically, this negation is an example of the amazing power of the mind, because in this case it is the mind that becomes blind to itself, and thus becomes incapable of accepting its very existence because it will not admit that it is made of matter!

The Idea of Subject

We come back to the idea, already apparent in the notions of *computo* and *cogito*, that everything that concerns mind or psyche is incomprehensible without the notion of subject (and that is why the notions of mind, psyche and subject are generally rejected all at once).

Certainly, Lacan's successors thought they could preserve the psyche by excluding the subject, and by replacing it with a sort of anonymity (The "one," the "it"), but they lost the very reality of the psyche. "If our brain is a television set, who is watching it?" asked Crick. We obviously need a reflexive duality, which is authorized by the ego circuit constituting the "I am me," so that the mind can recognize itself individually as psyche, and consciousness can develop itself by reflectively considering its own cognition.

Thus we can now envisage the organ, brain, neuro-cerebral equipment, the emerging qualities of the intelligence, the mind, and the psyche as many instances and moments of a same organizing, recursive, complex reality, which only becomes substantial through its own activity.

Conclusions

1. We can and must from now on reintegrate the mind with the *physis* (on this term, cf. Method 1) and the *physis* with the mind. We can even, in the same movement, reintegrate the mind with the *bios* and the *bios* with the mind.[3]

But to do this, we must stop thinking on the basis of the paradigm of simplification (disjunction or reduction), which can only separate the two terms or submerge the one in the other.

2. We have made the stupefying discovery that billions of billions of interactions among thirty billion neurons and a million billion synapses make up *one* mind, *one* thought, *one* judgment, *one* will.

That is because they are integrated/integrators in a recursive dynamism, that of the brain-mind, which is, at the same time, that of the *computo-cogito*, which constructs a being-egocentric subject. Kant spoke of the unification (synthesis) achieved by the judgment of the thinking subject, to which it must be added that the unifying act is a *computo-cogito*. The unifying *cogito* produces at the same time the knowing *computo* subject, which transcends itself by the very movement of its production. Let us remember, however, that this movement of production/organization can in no way be reduced to the category of the individual-subject, which can only be conceived of as understood in the circle of auto-(geno-pheno-ego-socio-)eco-reorganization. This circle is logically basic for conceiving the notion of individual-subject and becomes logically basic for conceiving and integrating the notions of body, brain, mind, and psyche, because it must be absolutely understood that the thing that knows is not a brain, is not a mind, but is a being-subject that knows through means of the mind-brain. "A human being is a human being, neither an observer hidden away in his own sensorium, nor a brain with arms" (S. Toulmin).

Everything that concerns the being, however, concerns the mind/brain, and everything that concerns the mind/brain concerns the being. "The mind that animates action is animated by the action of the whole being" (Method 2, 290). The spiritual processes need the cerebral processes, which in turn need the physiological processes, the body's machine regulating blood pressure, the rhythm of the heart, gastro-intestinal secretions, which in turn are controlled by the neuro-vegetative system, which is regulated by the neurocerebral equipment, which. . .

"A human being creates itself and recreates itself in a self-starting process of animation/corporalization. The mind is neither the renter nor the owner of the body. The body is neither the hardware, not the servant of the mind. They are both constituent elements of an individual being endowed with the quality of subject." (Method 2) This seemingly substantial body is only the concrete form of computing interactions of a being-machine that produces/organizes/regulates itself.

Thus the mind-brain is reintegrated into the whole body, but we must also reintegrate the human being within the society, which allows the fact-collecting of his brain to develop into thinking, via the language and the learning that is stored there. Without culture, the human mind could not reach thinking and would be stuck at the data-processing level of the lowest class of primates.

We can thus see that the problem of knowledge does not have only one unique source, and that we have an inseparable complex in us which, at each moment, in its own way, contains many parts:

BRAIN-------------------MIND

INDIVIDUAL

CULTURE-SOCIETY

Thus we can reintegrate the mind-brain with humanity and reintegrate humanity in the animal realm it surpasses, but contains and conserves. The humanity of knowledge is actually the union of the animality of knowledge and the humanity of knowledge.

Notes

1. As José Delgado states: "Without a brain, the mind cannot exist. Without sensory inputs, the mind will not be structured and cannot appear. Without manifestations of inner perceptions, and outward motor expression, the mind will not be recognized by the individual or the environment."
2. If mental representation is an imaginal configuration attached to an *ad hoc* interneuronal configuration, then we could conceive of a reciprocal effect from one to the other, the same as in hypnotic or magic suggestion wherein the imaginal configuration introduced from outside can determine the interneuronal configuration on which the psychosomatic processes of modification of the organism's states (by placebo, magic, witchcraft, etc.) would intervene. This would allow us to think that belief, or an idea, could modify a being.

3. "The living mind. The word living is of capital importance. The grandiose inscription written on the stone facade of the University of Heidelberg, *Am lebendige Geist*, salutes . . . , in the interpretation proposed here, no longer the immaterial substance descended from on high to inspire the master-thinkers, but the mind born from life and which can only unfold through becoming more and more alive. The living mind does not withdraw from life; it develops life in itself, in us. The human mind is no longer isolated far from life, but develops . . . a new life—the life of the mind in a new world, the world of the mind" (Method 2, 290).

References

Bateson, G. 1972. *Steps to an Ecology of Mind.* New York: Chandler.
Bourguignon, A. 1981. Fondements neurobiologiques pour une théorie de la psychopathologie. *Psychiatrie de l'enfant* XXIV.
Morin, E. 1981. *La méthode 2 la vie de la vie.* Paris: Seuil.

2.
Biological Bases of Real and Imaginary

José M.R. Delgado

Introduction

The concepts of "real" and "imaginary" are classical in philosophy and have been debated from pre-Aristotelian times to the present, often with a dualism similar to the controversies of mind and body.

In the present paper I would like to analyze the real and imaginary from a biological point of view. This approach avoids semantic fencing and has the advantage of introducing experimental testing without doctrinal positions. In this way, ideas may develop to conform with the growing knowledge and new data provided by rapidly expanding neurobiological sciences. Real and imaginary concepts depend to some extent on the structuring, evolution, and manifestations of cerebral activities. The universe provides a substratum of matergon (matter – energy) and small parts of it impinge on individual sensory receptors, to be codified as *transmaterial entities* requiring a *material carrier* and a pattern of *information* which is transmitted through neuronal networks to be individually processed, perceived, and eventually stored in memory, shaping individual personality.

The substratum is the universe. The meaning is in the brain. The relations between these entities have material aspects and transmaterial properties which can be investigated.

Both must be taken into consideration for a proper understanding of reality and imaginary, with the limitation that this understanding is also a function of the living brain.

Concepts may be considered as necessary tools for the brain so that it can process the codes and meanings provided by the senses. It is therefore convenient to analyze the essential, successive steps and feedbacks which link the universe to the individual perception and structuring of mental activities necessary for our inner understanding of the outer world. For this analysis I propose to consider the neurophysiology and neuropathology of the real and the imaginary.

Definitions

Proper definitions of concepts such as real, mind, and body involve many controversial issues and unknowns, making these definitions difficult or impossible for general acceptance. A better understanding of some of these concepts is precisely the issue of the present meeting. The attempt to define controversial terms in a few sentences may appear superficial and presumptuous, but to read a book about mind, for example, without understanding the basis of the author's orientation invites confusion.

In spite of their inherent difficulties, provisional definitions are convenient for a discussion of proposed theses. A definition may be unsatisfactory for the reader but it will clarify the position of its proponent, which may then be accepted or rejected. Having stated these limitations, I will define the meaning of five essential terms used in the present article: brain, mind, spirit, real and imaginary.

1. *The Brain* or cerebrum is a material entity located inside the skull which may be inspected, touched, weighed, and handled. It is composed of chemicals, enzymes, and elements that can be analyzed. Its structure is characterized by neurons, pathways, and synapses which can be investigated. The dead brain can be preserved in formaline and is the subject of anatomical, histological, and chemical research. In order to function, the brain must be alive, consuming oxygen, exchanging chemicals, and providing electrical activity. Many of the brain's chemical, thermal, and electrical phenomena subserve physiological activities unrelated to mental functions. The

brain is the material substratum of mental activities but should not be identified with them: the two terms brain and mind are *not* interchangeable.

2. *The Mind* is the term used for the ill-defined group of mental activities. Its existence requires a functioning brain. Without brain there is no mind, whereas the reverse is not true. For example, during deep anesthesia mental activities may be blocked while several brain functions such as regulation of respiration may continue at normal physiological levels.

The main problem in understanding the mind is that in layman's language as well as in philosophical and scientific discussions, the mind is considered as a *unit*. In reality, however, there are many kinds of mental activities: talking, enjoying music, and problem solving have different sensory inputs, neurophysiological mechanisms, feelings, and behavioral manifestations. Agreement will be difficult if authors assume that they are all referring to the same subject—the mind—when one is considering free will; another, consciousness; and still another, creative writing.

The mind is a dynamic entity, which cannot be frozen or preserved in formaline. Classical questions such as "What is Mind" or "Where is Mind" are, in my opinion, incorrectly stated because of the complexity and heterogeneity of mental activities. It is preferable to investigate and discuss the particular aspects and mechanisms of each mental function: for example, to determine those which play a role in visual recognition. The attempt to evaluate the many and heterogeneous mental activities one by one simplifies the problem, although the task is still tremendously complex.

In spite of the above-mentioned limitations, it is often convenient to talk about the mind in general terms. Mental phenomena are traditionally divided into cognition, affection, and volition. Epistemology, or the theory of knowledge, is one of the major aspects of philosophy and includes processes of attention, sense perception, understanding, memory, inferring and doubting. Examples of this are that musical tones such as middle C do not exist in the air but are qualities that mind itself generates when the appropriate hair cells in the organ of Corti are stimulated. Under certain circumstances, heat is felt as cold.

The nature of cognition and the relationships between the *knowing mind* and *external reality* have been subjects of philosophical debate since ancient times, reaching a peak with the "cogito, ergo sum" of Descartes and at present with growing interest in neurophysiological research on cognitive mechanisms that involve sensory perception, awareness, and judgment. My proposed *transmaterial* brain-mind theory, called "Triunism" (Delgado, 1979), postulates that the mind is a unity of three structural elements so essential that the absence of any one will prevent mental existence. These elements are:

a) Brain cells and pathways possessing material and transmaterial properties.

b) A flow of transmaterial, environmental information, coded and transduced at the sensory receptors and forming an active part of the working brain through a modification of its anatomy and physiology.

c) Detectable manifestations derived from (a) and (b) which are expressed inward as perceptions and outward as behavior.

3. *The Spirit*. A relationship between man and supenatural life is traditionally implied in the use of the words "soul," "anima," and "spirit." Unfortunately, the term "spirit" is commonly applied with a wide range of meanings from alcoholic beverages to disembodied souls, as well as referring to the third person of the Christian trinity. Another complication is that "Holy Ghost" is "Saint Spirit" in French and "Espiritu Santo" in Spanish, rendering the words "ghost" and "spirit" synonymous. The meaning of the term "spirit" is therefore so varied and confused that it will be little used in the present article.

We should, however, retain the word "soul" as a metaphysical concept of an entity incorporeal, immortal, having possibilities for salvation or damnation, and related to faith and theological considerations which cannot be proved or disproved with present experimental methods. The term "soul" is clearly different from mind. Science has reached the mind as a subject of experimental investigation while the soul remains outside the realm of scientific inquiry, being specifically related to metaphysical interpretations of human existence.

4. *Real*. As indicated by Charon, most physicists still conform to the reductionist idea that "there exists a concrete

Reality out there,'' and it is this ''concrete'' reality that mind tries to represent with increasing approximation.

Platonic concepts of specific and generic natures and the essential Aristotelian role played by the universal in reasoning led to counterparts in human thinking about what really exists independent of human existence.

In its evolution, realism, rather than being a doctrine stating the independent existence of things, is an evaluation of *cognition*. The external origin of sensations in Descartes and Locke was accepted as philosophical, common sense realism. The object in cognition and in reality, however, may be numerically two at the time of perception, according to the monism of the neo-realism of George Santayana and other philosophers. In direct realism, no intermediate objects exist between cognition and the external thing perceived. In representative realism and in the physiological idealism of contemporary neurophysiologists, the image directly apprehended by the senses (the sense-datum) is markedly different from the physical object.

In the present paper I will minimize philosophical discussions and, accepting the reality of the material universe independent of human existence, I will evaluate some of the neurobiological mechanisms of *cognition* in order to derive theoretical and practical consequences for the individual and for society.

5. *Imaginary*. Imagination is defined as the picturing power of the mind, a constructive faculty recombining mental images with depth and creativeness, seeking the essential truth. It is different from fancy, which is more superficial and playful, and from fantasy, which is more capricious and erratic. In the dictionary imaginary has the connotation of existing only in the imagination, being unreal.

Sensory reality requires a direct physical and temporal relation with the source of stimuli. Recall is the bringing to consciousness of a real experience from the past. The quality ''imaginary'' is the perception in the conscious mind of events which have *no relation with past or present reality*. This perception may be accompanied by the awareness of being unreal or, in the case of hallucinations, may be interpreted as being real.

I emphasize that the above definitions are not offered with the pretension of convincing the reader, and are only working tools to facilitate comprehension of the material presented in this article.

Neurobiology of Real

1. *The Universe as the Source of Stimuli*: To neurobiologists, the semantic and conceptual fencing of philosophers concerning imaginary and real is outside the realm of their experimental research. Their purpose is to understand the transmission of information from the external environment to the central nervous system. The external world is represented within the organism by patterns of neuronal activity carrying the internal coding of external information.

Stimulus is defined as a pattern of physical energy which activates specific sensory receptors, initiating the process of information coding. Stimuli can be identified, measured, characterized, and stored by means of instruments, independent of their possible biological effectiveness. For example, music can be stored on a tape and images can be kept on film or videotape.

Potential stimuli must be distinguished from actual stimuli. Physical energy must produce neural activity in order to be effective as a stimulus. This requires adequate sensitivity of the sensory receptor to the impinging energy. We lack sensors for many activities of the universe which therefore escape human awareness. Many previously unknown aspects of reality, ranging from ultramicroscopic elements up to the structure of the galaxies may be sensed and investigated using suitable modern detectors and amplifying instruments which provide sensory inputs of "unreal" pictures of the "real" world. Sensory receptors function as *filters* of information, and a large percentage of their activity is blocked peripherally without being transmitted to the brain.

The environmental *external* source of stimuli is essential for the activation of receptors. In total darkness, the eye will see nothing in spite of the normality of retina and visual pathways.

Stimuli from the environment are the *starting point* for a sequence of events leading to cognition. The presence of a stimulus, however, is not necessarily followed by a conscious perception because many other psychobiological elements are also necessary. The type of physical energy of the stimulus must be *adequate* for the particular sensitivity of the receptor. In the eye, for example, the visual receptors are tuned to a band of wave lengths from 400 to 700 nm.

Different types of energy may activate sensory receptors. For example, mechanical energy may activate the skin receptors or auditory cochlea. Accelerative forces, due in part to gravitation, are appropriate stimuli for the vestibular system. Other adequate stimuli are thermal, photonic, and chemical. Electricity is almost unique in its capacity to stimulate peripheral receptors, pathways, and central mechanisms of the brain.

One essential element in the stimulus is its *pattern*. In music, for example, a wide range of melodies may be structured with the same amount of acoustic energy. In animals as well as in humans, the psychophysiological effects of patterns may be very different (Lettvin et al. 1959, Segunda et al. 1963).

Sensory deprivation will have catastrophic consequences for the structuring and functioning of the mind. In the absence of stimuli, mental functions will not appear, or if already present, they will be totally disturbed.

2. *Transduction and Encoding of Stimuli*: The energy of the stimuli activating a specific receptor will produce a chain of neurophysiological events, starting in general with a local chemical or electrical disturbance and the triggering of electrical potentials.

In the *transduction* there is a conversion of energy into transmittable neuronal signals. The amounts of necessary energy are sometimes incredibly small: a single quantum of light is capable of activating a single human photoreceptor. In the *coding*, some specific pattern of the received signal elicits patterning of the neuronal discharge which conveys information. One simple example is the stretch receptor of the river crab, in which the frequency of spike discharges is proportional to the tension of the muscle.

Transduction is not an automatic process because sensitivity of the receptor may be modified by *feedback* from the brain. There is also a process of *adaptation* and a persisting stimulus could be less effective through time. This is probably one reason for the small irregular movements of the eye, which are continuously changing the position of the receptors used in normal vision.

The initial generator potential is graded and mediated through synaptic actions, determining an initial encoding of the received signals and leading to a propagated spike action potential. Receptor potentials are usually excitatory but may sometimes be inhibitory.

The high sensitivity of receptors to specific stimulus energy (photonic for the eye, acoustic for the ear, etc.) provides an initial differential sensory encoding. According to coding theory, there are invariants of organization, pattern or meaning which can be transmitted from original sources to final destinations independently of the symbols, kinds of physical energy, transmitting pathways, and intermediate processes involved.

A code may be defined as "a set of symbols that can be used to represent messages, concepts, and meanings (patterns of organization) and the set of rules that governs the selection and use of these symbols" (Uttal 1973).

An important controversial issue is whether *cognition* of the code requires a central decoder, in Bullock's (1961) words, to determine "who reads ensemble codes." No homunculus, no separate nonneuronal mechanisms need to be involved, but we must know how neuronal mechanisms are established for our individual understanding of the codes. Previous learning and experience are necessary to permit deciphering of meaning, which is therefore a learned ability and not a genetically established skill. Languages provide a classical example of coding for the exchange and processing of information. Languages can be translated but we need to know the code in order to understand the message.

The study of neural coding of information is basic for the understanding of transmission of signals. Perkel and Bullock (1968) proposed the "*candidate neural codes*," defined as any neuronal signal that varies as a function of some stimulus variable. Qualities of the received stimuli are encoded into transmittable electrophysiological events which propagate with the "all or nothing law," precluding the use of amplitude of spike modulation and insuring greater fidelity in transmission. Neuronal coding is poorly understood, but it may include modulation of frequency, bursts, interspike intervals, frequency patterning, macro- and microfluctuations of frequency, and other mechanisms. Each pattern of received stimuli conveying information elicits another specific pattern of electrophysiological events which carries an analog of the received message.

3. *Material Carriers and Transmaterial Information*: Any stimulus received by a sensory receptor originates in a *material* source located in the environment and functions as an emitter of patterned energy. Any activation of a sensory receptor requires transduction of energy and coding of external information as internal neurophysiological events. Between the peripheral senses and the central nervous system there are complex links of receiving cells, synaptic contacts, pathways, interconnecting neurons, dentritic arborization, somas, axons, and specialized neurons with processes of subcoding, decoding, storage, recall spatiotemporal comparisons, cognition, feelings, and outputs.

In the attempt to understand the transmission of information, we may consider a simple example: the visual perception of a red cross painted on white paper. A pattern of photonic energy will go through the air and the eye chambers, initiating specific chemical reactions in the retina. The chemical code is then transformed into an electrical code at the cones and rods, initiating a series of codified messages which will be transmitted through the optic nerve, optic chiasm, lateral geniculate bodies, and superior colliculus, reaching the columnar organization of the occipital cortex, activating memory storage of past experiences, and firing some neuronal structures related to awareness, cognition, and feelings. In these processes we should distinguish between two different elements:

a) *Material carriers* have been activated from the retina to the cerebral cortex, changing in location, structure, and time, and involving a succession of chemical reactions, enzymatic activities, release and re-uptake of neurotransmitters, depolarization of membranes, and other complicated processes. Glucose, oxygen, ATP, and other substances have been consumed. Microamounts of energy and heat have been liberated. These activities are the *material support* of neuronal functions.

Matter may be described as aggregates of elemental particles possessing inherent properties of energy, mass and inertia, occupying space, able to interact by impact, and capable of being destroyed, thus releasing energy. The existence and properties of material particles may be the subject of experimental research.

Material carriers of the nervous system are necessary for the transmission of information to the brain. In the absence of eyes, if the optic nerves are destroyed, or if glucose and oxygen are lacking, the red cross will not be perceived by the individual.

The material carriers may be changed without losing the meaning: the cross may be large or small, the image may be seen by the left or right eye, or an acoustic description may be visualized. The conceptual representation of the red cross will persist in spite of changes of carriers.

b) The second element in the brain activity is the *information*, or *message*, which should not be identified with the supporting material carrier. The information may be considered a *transmaterial* entity related to patterns of material organization or to temporal or functional relations between parts of the material substratum. Transmaterial entities may *transcend* the existence of their supporting materials, changing carriers while preserving the transmitted information. These entities do not possess material properties such as mass or energy. The same material amount of paper and ink may have many different patterns, meaning letters, words, and information. The message is the patterning of matter without adding or subtracting new material or energetic elements. The message may be erased without changing the amount of its supporting matter. Disintegration of matter is accompanied by liberation of energy, while messages can be destroyed without any energetic change. The concept of transmaterial is useful for the understanding of cerebral processing of information, and allows the experimental research on both entities: the material carriers with their neurophysiological aspects of neuronal functioning, and transmaterial information with its related processes of coding and decoding of patterns of transmitted signals.

4. *Circulation and Processing of Codes:* The visual system provides a good example for analysis of the circulation of coded signals in the central nervous system. The retina is a duplex formation with one type of rod and three types of cones. Using glass microelectrodes with very fine points (0.1

micron), Tomita et al. (1961) were able to record a graded DC hyperpolarizing potential when a cone was stimulated by light with peak sensitivity in three different classes of cells in the blue, yellow-green, and red portions of the visual spectrum. This may be considered the first stage of color coding at the photo-receptor cell. The next step is in the retinal, horizontal and bipolar layers which have a very broad spectral sensitivity, producing a graded hyperpolarizing response. The majority of cells respond with hyperpolarization in the blue and depolarization in the red. Other cells, probably bipolars, have a triphasic response, hyperpolarizing at both ends of the spectrum and depolarizing in the middle of spectral stimuli.

The important point is that *the coding schema for colors is changing at the second stage of neuronal processing.* At the third neuronal level, formed by the amacrine and ganglion cells, there is a new element in the coding scheme: the initiation and propagation of regenerative spike action potentials (Werblin and Dowling, 1969). This stage is important because the axons of the ganglion cells make up the optic nerve. Recordings in this nerve show two types of responses (Granit 1945): one with relatively narrow spectral sensitive curves, with center wave lengths at about 460, 540, and 600 nm (chromator modulators), related to the color receptors of the retina, and a second type with a very broad spectral range, almost 125 mm wide, referred to as scotopic and photopic dominators. The visual mechanism is further complicated by differential opponent spike rate responses, temporal "on" and "off" responses, and patterns of burst activity related to chromatic characteristics of the stimuli (Wagner et al. 1960).

The fourth level of neuronal processing is in the lateral geniculate body of the thalamus. Bursts of spike activity similar to those present in the ganglion cells may be recorded at this level (DeValois et al. 1966) with two types of responses. In one group of cells, steady firing was present and could be increased or decreased depending on the chromatic type of the stimuli (spectral opponent cells). A second group of cells uniformly increased or decreased its activity depending on the spectrum.

In addition, most cells in the lateral geniculate are affected by the shape and location of the stimuli (Hubel and Wiesel, 1979). At this level, therefore, there is an overlapping between spatial and chromatic neural coding.

The fifth stage of neuronal processing occurs in the visual cortex of the occipital lobe, where recording of single cells in monkeys has been performed by Motokawa et al. (1962), and other investigators. In this cortex, many cells responded with trichromatic sensitivity tuned with peaks at 460, 530, and 620 nm. In this way, the original trichromatic information is re-encoded in terms of opponent mechanisms, re-emerging at the cortical level with a trichromatic code.

The neurophysiological data presented briefly in this section provides some idea of the complexity of circulation of codes, emphasizing the existence of a chain of material carriers of information which may be investigated experimentally. The neurobiology of transmitting mechanisms gives a rigid limitation to the *codification of signals, which necessarily must be expressed as electrical patterns of neuronal firing.*

5. *The Meaning is in the Brain*: There is a contrast between the simplicity of some basic coding mechanisms and the power and complexity of the information which can be transmitted. The Morse code, for example, uses dots and dashes to transmit messages in any language to any corner of the world, and it can deal with any scientific, economic, historical or other kind of information. The computer is a digital instrument with a very simple coding of yes-no which, however, has a huge capacity to receive, transmit, process, and display any kind of data.

The nervous system also has a relatively simple general purpose mechanism for handling information: the sodium-potassium-chlorine nerve membrane model. This mechanism is rather similar in squids, frogs, cats, monkeys, and man.

Sensory systems have an initial discrimination compression and processing of information at the receptors, a more simplified transmission along the three or four interconnecting neuronal links, and then an extraordinary anatomical, chemical, and functional complexity at the cortical level, where the most important and delicate processing of information takes place. The understanding of information is based on unknown neuronal mechanisms located "somewhere" in the brain—and not in the heart or liver.

A red cross painted on paper may provide a source of stimuli crossing the dioptics of the eye, stimulating the retina, and triggering electrical discharges travelling through the optic nerve. With suitable electrodes in an anesthetized cat, for example, we may record the electrical patterns at different stages of the visual pathways related to the presentation of the visual image. We cannot expect to find a red cross in the bipolar cells, in the superior colliculus, or even in the neurons of the visual cortex. A microelectrode may record only the electrical events of one neuron, but the visual stimulus is activating thousands of neurons which are interesting with each other and with sequential events of unknown temporal characteristics. The technical difficulties of recording the many intervening variables are enormous. Suppose, however, that this impossible task were solved and that we had a huge computer filled with all recorded signals. The result would be a complex code used to digitalize the sensory input.

A different set of instructions would be needed to *extract the meaning* (a red cross) from its codified neuronal signals. Sensory pathways only transmit codes. The software code for decoding must be previously acquired for the understanding of the received information. A computer without *instructions* to classify and cross-correlate inputs and provide logical outputs is a useless piece of hardware. For a person to read a publication or understand a spoken word, previous learning of the codes of the specific language is necessary. Without this previous experience, a language is only noise without meaning.

A red cross, a book, or the related electrophysiological events evoked inside sensory pathways are only *material carriers* of information with "secret" messages waiting to be decoded by special neuronal mechanisms in order to have *real personal meaning*. A red cross may produce electrophysiological phenomena within the visual pathways of rats, dogs, and humans. The meaning depends on previous learning and animals may be taught that the cross means punishment or reward, depending on previous associations. For a human being, imagination may be triggered by memories of ambulances, flags, decorations, organization bracelets, and other symbols.

The meaning of symbols is not in their material carriers and is not genetically determined in the neurons: it must be learned by the brain, stored in memory, and brought to consciousness (awareness) by comparing present and past sensory codes.

The "reality" of the environment is necessarily coded at the sensory receptors, circulated, and then processed within the brain to produce a personal sensation of "real" according to frames of reference originating also in the world external to the individual, mainly in the cultural sets of values and knowledge. Perception of "real" and "imaginary" are individual intracerebral events linked to highly complex neuronal mechanisms which, although mostly unknown, may now be investigated experimentally.

6. *Frames of Reference to Understand Reality*: As indicated by Prechtl (1969), the infant's brain is not a miniature model of the adult brain, but a relatively simple nervous system with particular properties at each phase of its development. These properties are determined by genetically fixed programs of growth and also by environmental factors, most of them received through sensory inputs which influence both the function and structure of the central nervous system.

In the foetus, functional cerebral development proceeds from the reticular core with an initially poor flow of information to peripheral areas of the brain that have a high capacity for information flow. Unit activity in the rabbit's cerebral cortex appears between the fifth and tenth post-natal days (Schade 1960). Foetal reflex sequences and perinatal and early postnatal reactions are automatic in nature (Humphrey 1969). Immediately after birth, the baby is a competent learning mechanism (Lipsitt 1969) able to process information and to be conditioned in tests of sucking behavior, head turning, suppression of ongoing behavior, and other responses. One important issue is "to what degree is the naive organism at birth already prepared to organize and respond adaptively to the objects in the world before it has 'learned' about those objects? . . . Euclidian rather than another geometry might, for instance, be man's biologically 'given' way of organizing space" (Wolff 1969).

Anticipatory morphological maturation exists quiescent in the foetus: oral suction, respiration, kidney secretion, and gastro-intestinal activity are ready to act several weeks before the nine-month term is completed and may be triggered if the baby is born prematurely. No comparable provisions exist for mental functions. The newborn is able to learn but unable to speak or to comprehend the loving phrases of his mother. He has no words, concepts, tools for communication or culture. Lacking previous information, there are *no frames of reference to understand the surrounding reality*. Early in life, the infant is attracted to sources of comfort and repelled by sources of distress. These experiences lead to association of sensory inputs and to progressive learning to recognize objects, persons and their own movements. In mammals, about 80–90% of the neurons of the hippocampus, cerebellar cortex, and other structures form only *after birth*. Experience provided by sensory inputs influences the number and structural connections of many post-natal cells. According to Cajal (1901–1911), the microneurons of the cerebellum develop after birth under the influence of the infant's behavioral activities. The ability to perceive and recognize visual patterns does not pre-exist in the brain nor is it genetically determined: it must be learned by experience. In man, the sensory experiences during the early post-natal period are of decisive importance for material structuring of the brain, for the acquisition of cerebral mechanisms, and for the symbolization, perception, and comprehension of sensory stimuli.

The newborn brain has the potential to learn but not the reality of acquired information and experience. The *potentiality* to learn and use languages depends on genetic factors which exist in man and are lacking in animals. The use of specific languages requires *individual learning*. The baby does not invent nor can he choose to learn English or Chinese. The infant brain is entirely dependent on informational codes of reference which must come from the outside, from culture, and must penetrate the brain through the sensory portals of entry.

One important point is that because of the lack of brain maturity and the initial absence of mental functions, the first group of crucial choices of environment, inputs, and information cannot be made by the baby but must be selected by the adults in charge of his development and education. This decisive responsibility is therefore not individual but social.

7. *Shaping the Personal Mind*: A computer may be programmed in Basic or in other more complex languages. This is the decision of the human programmer and not of the computer. Later, highly sophisticated software may be given to the machine for creative work, including the development of new

programs, resolution of scientific problems, artistic productivity, and other "intelligent" tasks. Without programs and information, the computer cannot function.

A newborn human brain may be programmed with sets of associations (conditioning), values, languages, faith/religion, emotional reactivity, motor skills, knowledge, and many other elements. It may also be programmed with critical evaluation, personal dignity, ideological freedom, and other qualities. What is *not possible* is to leave the baby in sensory isolation because it would be highly detrimental for the anatomy and physiology of his brain and also for his behavioral reactivity. Nor is it possible to give the newborn free choice of his meals, physical care, or sensory inputs: food, personal care, affection, and information must be provided by adults. We must realize that within some limits of genetic preferences, the culture and care of the baby will have a decisive influence on his personality.

There is a great difference between brains and computers: the neuron is not a digital element but a highly sophisticated and complex unit. The information received through sensory inputs will influence dendritic growth, cellular structure, and neuronal chemistry, shaping brain structures materially. In contrast, the hardware of the computer is not changed by the software of inputs. Pattern detection seems to be related to specific neurons of the cerebral cortex which are shaped in their specific sensitivity by early individual experience. Experiments with kittens have shown that during a critical period of 3 to 14 weeks after birth, daily exposure for one hour to an environment with vertical stripes resulted in a specificity of response of most cortical neurons for the stimulus pattern received. *Information-extracting neurons* have been described as able to respond to universal features such as temporal and spatial transients, or to species-specific features, for example bug detectors in the frog, rotation movement in the fly, and color-sensitive cells in the monkey (see summary by Blakemore 1974).

In the auditory system there are also feature-sensitive neurons that abstract information from complex auditory environments. There are cortical neurons capable of differentiating noises, clicks, tones, and species-specific calls. Some cortical neurons may "respond" to questions such as "Is the stimulus on?", "Has it just commenced?", "Is the frequency changing?", "If so, in which direction and at what rate?", "Where is the stimulus located in space?", "Is it moving?", and so on (see summary by Evans 1974).

The complex circuitry necessary for some of these functions does not seem to be preprogrammed by genetic determinants and apparently depends on environmental information impinging on the developing sensory cortex. In this way, the received information shapes the functional characteristics of the neuronal systems.

Initially the brain and the sources of messages have independent existences. Codes, symbols, and meanings are not intrinsic—or emergent—properties of matter but intellectual agreements of human minds, based on invented systems of reference, learned individually through personal experience, and transcending personal existence. The alphabet was invented long ago and will be used for centuries to come.

The reverse phenomenon of cerebral materialization of symbols takes place during verbal expression. In this case, the material substratum in the neurons activates the codification and transduction of symbolic representations, producing at the end the vibration of vocal cords and production of sound waves carrying transmaterial messages. All these processes are unconscious and automatic.

Symbols may be shared by human beings and animals. Cats and monkeys can learn the meaning of many stimuli. Development of personality is primarily the experiential accumulation of symbols and frames of reference for decoding sensory information. The usefulness of the concept of transmaterial is that we must deal with it as a research tool when we explore the meaning of matter.

The personal mind has been considered inviolable, free, beyond the control of governments, and owned only by the individual. Recent advances in technology, world organization, and psychophysiology have modified these concepts. The individual does not control his genes, liver function, or the synaptic linkages of his neurons. Most essential aspects of our biology are beyond personal awareness and control. In addition, living in societies under innumerable restrictions, we are obliged to speak a specific language, to conform to the social norms of our group, to submit to structured education for a certain number of years, and to obey sets of laws and regulations. What is more crucial, throughout this developmental period a variety of values, desires, ethics, and information have been introduced into our brains without our consent, awareness, or understanding. Being educated and immersed in the culture of the Soviet Union, the United States, or Iran is more decisive for the structuring of the personal mind than initial genetic endowment.

The conclusion we are reaching is that the personal mind is formed by sets of coded sensory inputs which originate in the environment, creating frames of reference which are decisive for the interpretation of reality. Fortunately these sets can be partially modified by intelligent evaluation of available information.

Neurobiology of Imaginary

1. *Reality as a Necessary Source of Imaginary*: The unicorn is a mythical animal resembling a horse and first portrayed on Assyrian reliefs. It has a white body, a purple head, blue eyes, and a single horn with a red pointed tip. As a biblical animal, the unicorn became the object of allegorical interpretations and was often linked to Christ, who raised up a horn of salvation for mankind. Hunting of the unicorn was represented in medieval art in Europe, the Islamic world, and China.

Could the unicorn be created in the absence of horses, rhinoceroses, and without the poetic interpretation of human beings? Probably not. The World of the imaginary, populated by mythological gods, ghosts, fair ladies, demons, and a wide range of exotic animals was a caricature of the real existence of man and animals. A different imaginary world has recently been created with intelligent apes, inhabitants of lost planets, wars of galaxies, and space ships. These fantasies are based on the reality of apes, computers, and outer space technology.

The human mind lacks the power to create images from nothing but it is able to combine bits of information and to escape accepted rules of reality. A fiction novel or a film deals with imaginary characters who amuse or move us because of the close relations with the emotional aspects of the real world. The brain, like the computer, requires previous information in order to function. For a person born blind, it is difficult to imagine the colorful beauty of a landscape. For a deaf person, music has no meaning. Invention is the discovery of new, unknown associations. Creativity requires materials and experience to give shape to a painting or a building or to organize an industrial plant.

Imagination is necessary in human life, but a distinction between real and imaginary is essential. Under the effects of a psychoactive drug, some people have believed they have a sudden power to fly and have jumped through a window, with obviously disastrous consequences.

The personal perception of a sensation—or an idea—as real or imaginary depends on mental activities and on the processing of information. These phenomena can be studied because they have a real existence as neurophysiological events.

2. *Essential Neuronal Mechanisms*: Under deep anesthesia, mental activities are blocked and there is no perception, volition, or thought. Several areas of the brain can be destroyed without causing notable impairment of mental functions. Psychosurgery and therapeutic interventions for tumors, epilepsy, and behavioral disturbances have ablated or disconnected the frontal lobes, cingulate gyrus, parts of the limbic system, and other cerebral areas without altering the capacity to imagine. In contrast, small tumors or lesions in the hypothalamus or reticular formation may induce severe impairment of consciousness. Electrical stimulation of the human brain has been able to evoke imaginary perceptions only when specific areas within the temporal lobe, occipital cortex, and a few other regions are activated. Cortical receiving areas have been identified for the senses.

Localization theories are rather controversial, and even in the motor cortex the representation of an inverted homunculus is too schematic. Motor activities require the harmonic combination of many structures including the cerebellum, basal ganglia, and cerebral cortex. The multianatomical participation of different areas of the brain in mental manifestations also seems to be necessary. Investigations may be performed to determine whether a specific structure is essential for a specific mental performance. For example, speech is related with the integrity of limited areas of the temporal lobe. Perception of sensory inputs as real or imaginary requires consciousness and therefore integrity of the reticular formation, but at present it cannot be associated with specific cerebral areas, although results of electrical stimulation suggest that the temporal cortex may be involved.

3. *Triggering of Imaginary Perceptions*: Perception of sounds, images, and past experiences may be triggered easily by exercising the faculty of remembering. Some people can voluntarily recall vivid pictures of the past which may be perceived with qualities close to reality, while they remain aware that these images are only in memory. Dreams may produce an extraordinary amount of visual, auditory, and other sensory phenomena. These perceptions, however, are not included in our provisional definition of imaginary which has the connotation of existing only in the personal mind.

Imaginary peceptions may be triggered artificially by electrical stimulation of the brain. The primary visual receiving areas, located in the occipital cortex, may be activated by electrodes implanted in the occipital cortex (Brindley 1982) or even transdermally through the intact scalp (Merton and Morton 1980). Luminous flashes (phosphenes) evoked by electrical pulses are an example of sensory perception of nonexisting light. Electrical activation of the temporal lobe, as described by several investigators (Delgado 1969, Penfield and Jasper 1954), has been able to evoke the perception of smells, specific musical melodies, voices, scenes with people talking and behaving, and other phenomena which were often interpreted by the patient as completely real.

The onset of some seizures may be accompanied by well identified sensory experiences. Neurobiological mechanisms are essential for the perception of real and imaginary, as shown by the well known effects of psychopharmacological agents. Hallucinogens may produce changes in perception ranging from sensory illusions, meaning distortions of what is perceived, up to hallucinations with the perception of nonexisting objects. The action of hallucinogens is related to the activity of cerebral neurotransmitters, including sympathomimetic amines (catecholamines), indoles (serotonin), acetylcholine, and other compounds. The world of imaginary perceptions is therefore dependent on the processing of intracerebral codes related to chemical, anatomical, and physiological phenomena which can be investigated and to a great extent can be identified and understood.

Neuropathology of Real and Imaginary

Disturbances in any of the successive steps which intervene in the origin, reception, coding, transmission, processing, decoding, re-encoding, and final activation of conscious perception of information may block, distort, or falsify perceptions and meanings. The absence of information itself is extremely detrimental for development of the nervous sytem. One of the problems of low income groups and underdeveloped nations is the cultural deprivation—the lack of inputs. Sensory poverty represents an absence of stimulation for neuronal development, and it produces irreversible harm, failing to promote the wealth of genetic potential in the infantile brain.

Another problem is the manipulation of information and education given the child, which will be decisive for the initial structuring of his individual frames of reference and therefore for personal identity, ideological structure, and behavioral reactivity. The neurophysiological mechanisms involved in interpretation of sensory inputs can be disturbed in many ways. A myopic eye will provide a blurred view of the surroundings unless optically corrected. What is more important, neuronal activities depend on many complex physiological functions. Metabolic disturbances may originate in the pancreas, the liver, or in other organs, and they will have important consequences for brain function. Schizophrenia may be related to alterations in the synthesis and release of catecholamines. Depression and mania may depend on intracerebral neurochemical alterations.

In addition to genetic and psychological reasons, the problems of mental illness, the histology, neurophysiology, and chemistry of neurons are decisive for the normality and abnormality of the mind. We must therefore consider the neurobiological aspects in an attempt to understand the philosophical interpretations of mental functions.

Reality as Mind

The neurobiological considerations presented in this article support the concept that *reality* may be considered the source of stimuli which are coded, processed, and stored by the brain, providing sub-codes which are perceived by some unknown neuronal mechanisms as personal sensations. For the individual, *reality is only in his mind.*

In agreement with our transmaterial theory of the mind, in the absence of sensory inputs (information and coding), the brain (processing and frame of reference), and outputs (feelings and behavior), mental functions cannot exist and therefore reality cannot be perceived by the individual.

An understanding of neurobiology is essential for the comprehension of all human creations, including language, the conception of the world and mind, and interpretations of real and imaginary. Instead of debating philosophical positions, it would be preferable to recognize the limitations—and possibilities—of the human intellect. The unity of the universe must be divided into small parcels because of the limitation of our brain, which can receive inputs at only a relatively slow pace and in small quantities. Monism or dualism are only secondary problems of interpretation. The main question is to understand brain functions and their consequences for our relations with the environment.

In summary, instead of a rigid conception of reality, it is preferable to provide conceptual options, clarifying the personal and social consequences of each option.

Imaginary as a Source of Reality

If we define imaginary as inner mental perceptions unconnected with incoming sensory inputs, then it is not acceptable that the imaginary could be a source of the real. We may, however, consider the opposite view, because external reality is coded at the sensory inputs and must be compared with the stored frames of reference in order to be understood. Thus reality determines a neuronal processing of codes, originating new subcodes which are perceived as reality. Coded signals and neurological processes are different from physical reality. They are only in the mind. In this context, we may consider the imaginary as a source of reality. Rather than defend doctrinal positions, it is more important to understand the interpretations and to have a clear idea of the biological processes which determine perceptions and feelings.

Absolute Values: The Scientific and Moral Truths

The established frames of reference and the setting of values will determine the interpretation of present reality. Interpretations and values depend on cerebral processing of signals and comparison of the information stored in memory with the incoming inputs. It is therefore difficult to accept that anyone could possess absolute values or truths, independent of the messages received from the environment.

What is possible and advisable is for society to establish sets of values for individual and social well-being and development. These sets must be provided during early education. One important value to be taught is a sense of personal freedom that will permit critical, intelligent evaluation of the cultural system. Personal freedom and democratization of science will increase individual participation in the future evolution of mankind. At the same time it will provide a greater sense of human dignity and personal accomplishment.

The establishment of sets of values must have universal characteristics and must be acceptable to most people regardless of their race, sex, religion, political ideas, and other differences. This is possible because all human beings share a similar biology in their bodies and in their brains. Basic biological needs are also similar in all people. The establishment of educational programs and life goals requires a general plan that must be multinational and multidisciplinary. A biological orientation of the future evolution of mankind could avoid aggressive trends and could be an instrument for peaceful coexistence. This effort deserves the attention and collaboration of national and international agencies.

References

Blakemore, C. 1974. Developmental factors in the formation of feature-extracting neurons. *The Neuro-Sciences: Third Study Program*, edited by F.O. Schmitt and F.G. Worden. Cambridge: MIT Press.

Brindley, G.S. 1983. Effects of electrical stimulation of the visual cortex. *Human Neurobiology* 1:281–283.

Bullock, T.H. 1961. On the anatomy of the giant neurons of the visceral ganglion of *Aplysia*. *Nervous Inhibition*, edited by E. Florey. New York: Pergamon Press.

Cajal, S. Ramon y. 1909. *Histologie du systeme nerveux de L'homme et des vertebres*, volume I and II. Paris: Malorie.

Delgado, J.M.R. 1969. Physical control of the mind: *Toward a Psycho-Civilized Society*, volume XLI, World Perspective Series. New York: Harper & Row.

Delgado, J.M.R. 1979. Triunism: a transmaterial brain-mind theory. *Brain & Mind*, Ciba Foundation Symposium 69 (New Series). Amsterdam: Excerpta Medica.

DeValois, R.L.; Abramov, I; and Jacobs, G.H. 1966. Analysis of response patterns of LGN cells. *Journal of Optical Social America* 56: 966–977.

Evans, E.F. 1974. Neural processes for the detection of acoustic patterns and for sound localization. *The Neuro-Sciences: Third Study Program*, edited by F.O. Schmitt and F.G. Worden. Cambridge: MIT Press.

Granit, R. 1945. The colour receptors of the mammalian retina. *Journal of Neurophysiology*.

Hubel, D.H., and Wiesel, T.N. 1979. Brain mechanisms of vision. *Scientific America*, 24:150.

Humphrey, T. 1969. Postnatal repetition of human prenatal activity sequences with some suggestions of their neuroanatomical basis. *Brain and Early Behavior*, edited by R.J. Robinson. London, New York: Academic Press.

Lettvin, J.Y.; Maturana, H.R.; McCulloch, W.S.; and Pitts, W.H. 1959. What the frog's eye tells the frog's brain. *Processing Institute Radio of Engineering* 47:1940–1951.

Lipsitt, L.P. 1969. Learning capacities of the human infant. *Brain and Early Behavior*, edited by R.J. Robinson. London, New York: Academic Press.

Merton, P.A. and Morton, H.B. 1980. Stimulation of the cerebral cortex in the intact human subject. *Nature* 285:227.

Motokawa, K.; Taira, N.; and Okuda, J. 1962. Spectral responses of single units in the primate visual cortex. *Tohoku Journal of Experienced Medicine* 78:320–337.

Penfield, W. and Jasper, H. 1954. *Epilepsy and the Functional Anatomy of the Human Brain*. Boston: Little, Brown.

Perkel, D.H. and Bullock, T.H. 1968. Neural coding. *Neuroscience Research Program Bulletin* 6:221–347.

Prechtl, H.F.R. 1969. The problems for study. *Brain and Early Behaviour*, edited by R.J. Robinson. London, New York: Academic Press.

Schade, J.P. 1960. Origin of the spontaneous electrical activity of the cerebral cortex. *Recent Advances in Biological Psychiatry*. New York: Grune & Stratton.

Segundo, J.P.; Moore, G.P.; Stensaas, L.J.; and Bullock, T.H. 1963. Sensitivity of neurones in *Aplysia* to temporal pattern of arriving impulses. *Journal of Experienced Biology* 40:643–667.

Tomita, T.; Murakami, M; Hashimoto, Y.; and Sasaki, Y. 1961. Electrical activity of single neurons in the frog's retina. *The Visual System: Neurophysiology and Psychophysics*. Berlin: Springer.

Uttal, W.R. 1973. *The Psychobiology of Sensory Coding*. New York: Harper & Row.

Wagner, H.G.; MacNichol, Jr., E.F.; and Wolbarscht, M.L. 1960. The response properties of single gaglion cells in the goldfish retina. *Journal of General Physiology* 43:45–62.

Werblin, F.S. and Dowling, J.F. 1969. Organization of the retina of the mudpuppy. *Nectrurus maculosus*, volume II. Intracellular recording. *Journal of Neurophysiology* 32:339–355.

Wolff, P.H. 1969. A summing-up. *Brain and Early Behavior*, edited by R.J. Robinson. London, New York: Academic Press.

Discussion

José M.R. Delgado

Jerzy A. Wojciechowski: You spoke of the constant flow of thoughts which goes through our brain. And you mentioned that today we are not what we were yesterday, in the same way that tomorrow we will also be different. I would like at this point to remind you of a great brain specialist, very well known, whose name is Aristotle. Aristotle analyzed the phenomenon of change, and he states that in order to have change, it is necessary to have two things: something that is changing and something that does not change. In the same way, when you speak of the human being, it is true beyond a doubt that he is not exactly today what he was yesterday, and will not be the same tomorrow. But, however, there is always an "I". I was myself yesterday, I am myself today, and I will still be myself tomorrow. There is thus really something that stays the same in the midst of change, and this something is not at all abstract, but rather something very concrete. This concrete something is reflected in my consciousness, in my sense perceptions, in my personality and this is a very important point. There is something in me that never changes.

José Delgado: I disagree absolutely with your statement and, with all due respect to Aristotle, I think that we are continually changing. It is true that if you were a philosopher yesterday, you are still one today, and you will still be one tomorrow. But these are only labels which we can accept as not changing. But if you look at yourself, look for example at your blood, your blood changes; it only stays the same in your veins for about six weeks. After that, you make new blood. How long do your ideas last? Can you compare yourself with the child you were at three months, when you were three months old? You have not kept the slightest memory of what you were at that time.

Jerzy A. Wojciechowski: In spite of all that, however, I am still the same person.

José Delgado: No, absolutely not! I am sorry for this disagreement, but I think that what we are missing is a precise definition of what we call a human being. I am thinking like a scientist, and I am looking at how the mind and the human body function, and I am forced to conclude that we are continually changing.

Jean Charon: Jerzy does not mean that we do not change, but he is saying that there is something *at the very deepest level* of our being that does not change, which is precisely our "self". This is not necessarily something that shows up in a purely biological examination of the brain.

José Delgado: Yes, I agree. We must really define what it is we think is changing, and what does not change. From a biological point of view, I know that we are always changing.

Jack Kapenan: I am a psychology professor at the University of Miami, Florida. I congratulate José Delgado for his brilliant presentation of contemporary neuro-physiology. I see here used for the first time such "advanced" terms as transmaterial, or transcendental, which is not so usual in scientific circles. I wonder why, however, you suddenly become so cautious, and you avoid taking the next step. You do not seem to want to use one term especially, the term "transpersonal," which is the term used in psychology to mean whatever is related to the extension of the person, whatever integrates all the elements of the personality which are of a spiritual or creative nature. You were absolutely daring and full of imagination when it came to integrating many concepts; then, you suddenly came to a stop even though there are so many new concepts you should take into consideration at this stage, like the concept of *the Atman*, for example, which someone mentioned here. Why come to a halt, as if you were facing a great obstacle, when it comes to concepts which directly concern the mind? The brain, such as you study it on the level of neuro-physical terms, has, without any question, something to do with the deepest levels of the brain's activity, with what we psychologists call the mental, that is, the mind.

A second question concerns a number of "discoveries" (why not call them that?) made by parapsychology, such as the one, which now seems verifiable, of the possibility of seeing a certain number of images from far away. You are no doubt aware that many experiments and studies using scientific methodologies have been done on these "discoveries." Does this perhaps upset too "material" view of phenomena, as you presently view it in neurophysiology?

José Delgado: No, not at all! And I do understand your comments. But you know how cautious I was, and how I insisted on *defining* things so that I only talked here about what things can enter into a discussion of what we call science. For that very reason I tried to define as precisely as possible the terms brain, mental, and mind. These are unquestionably different concepts. You recognize this yourselves, and I think that we cannot treat a problem like that of the mind by simply referring to definitions which concern the brain or the mental process. One hundred years ago, the flight of the bat was cited to support the ideas of parapsychology. Now, with concepts and knowledge about things like radar, this phenomenon is no longer considered part of "parapsychology," but part of science. We are cautious not to use concepts that are only *partially* defined; and I am afraid that this is a great part of the attitude or approach to phenomena now classed as "parapsychology." When these facts become part of the scientific approach, then there will be no problem to look at them.

René Berger: You have always addressed us as human beings, and especially when you analyze the brain as being an organ unique to the human being. I have a question concerning this human being. Today, especially with the computer, the human being has an increasing interaction with the machine. I would like to know what you think about this interaction between humans and machines, and to what degree this could modify, if not the knowledge that we have about the brain, at least the functioning that we can expect from the human brain. We are only, of course, at the very first stages of this human-machine interaction, but this is going to develop and take off enormously in future generations. What is your feeling on this subject? For me, this seems fundamental. We are no longer only operating among human beings, but we are working among human beings and machines.

José Delgado: Nothing can penetrate our brain except that which has gone through our senses; and, conversely, whatever goes through our senses can profoundly alter the structure of our brain. Without question, then, we must take into account all that touches on the content of our brain and that concerns the interaction between humans and machines, as we have especially seen in the "ineractive" development between man and machine. Saint Thomas Aquinas and Leonardo da Vinci, among others, had already noted this inevitable relation between our senses and the content of our brain. This implied the relation between what our senses perceived and the content of our brain. At the present time, then, we can discuss those things that computers let us perceive, and the structure of our brain. More than twenty years ago I myself proposed and conducted experiments to study the relation between the content and functioning of a computer's memory and of the human brain, and also the relation of feed-back, that is, how the human brain itself could then make changes in the computer. That was a study of a real "coupling" of the computer and the human brain. Thus, I am in complete agreement with you. Technology, and specifically the computer, can greatly modify the content, and even the structure, of our brain. You are right.

A member of the audience: I am a medical student at Columbia University in New York. I think that what you have described calls for a methodology which you yourself have called a "scientific methodology." But you state that everything changes, and I think that methodology itself can change in order to develop more and more our knowledge about phenomena. It seems to me that biology is still using a very analytical and very cautious methodology that we were using in the last century and at the beginning of this century. But don't you think that we should look for some newer methodology, as we see now developing in physics, so that we can study phenomena like those you yourself mentioned, like these "transmaterial" signals which travel along the neurological circuits? Would not a new scientific methodology be necessary so we can progress, as, I think, Willis Harman will next discuss?

José Delgado: Yes, I think that you are right. In fact, in biology, too, we have begun to use more advanced methodologies, or new methodologies, especially where electromagnetism and the recording of electromagnetic phenomena associated with living things are concerned, as well as for electromagnetic stimulation of living structures. But we need not only a new methodology but also a new interpretation. I am a biologist, but I realize that more and more interpretations of phenomena must rely on philosophic concepts. That is why I am overjoyed to see this interaction between my biological concepts and the participants here, especially the philosophers. I as a biologist can present my paper to this gathering, and I am happy to see the way the participants from other disciplines react to my thoughts as a biologist. At the present state of our knowledge, we need a new methodology, and also a new interpretation, essentially an *interdisciplinary* interpretation of phenomena.

Mitsuo Ishikawa: Concerning the interpretation of electromagnetic phenomena, I myself proposed at last year's meeting in Fez, Morocco, the necessity of considering electromagnetic phenomena as "complex" in the mathematical sense of the word, so that we can interpret living phenomena more complete.

José Delgado: That is an interesting idea. What is essential to see is to what extent such an interpretation can bring together our knowledge and our present speculations with the facts drawn from experiments. We must always go back to experiments.

3.
The Real and the Imaginary in Complex Relativity

Jean E. Charon

Physics and representation

One of the central problems in contemporary theoretical physics is the interaction between an observed phenomenon and the consciousness of the observer considering the phenomenon. Since Descartes, we could say that physics has been accepting a clear distinction between the subject and the object, between mind and matter. The new idea which has emerged in physics today is that there takes place a direct interaction between subject and object. It is not only the idea that the subject, because of the limitations associated with his perception of the outside world, can obtain only a relative perception of this world (relative to the limitations of his senses). This has been obvious since Descartes himself. What is meant today is that mind is an "existing object of the world by itself, and that there might take place an interaction between mind and matter as between any two objects in this world, the interaction modifying the structure and the comportment of both objects." It is what the American physicist John A. Wheeler calls the direct "participation" of the subject in the phenomenon that he is presently observing.[1]

To understand this important new idea of modern physics it should first be clear that this approach to the world is not the result of a sudden interest of physicists in mind and consciousness, a subject that has long remained a "taboo" for the so called "exact" sciences. I personally believe that physics will pretty soon become an ontological physics, dealing with such "metaphysical" problems like *Being* (the absolute), *Spirit* (the soul) and the *Word* (space-time relations between all phenomena); and I have justified myself elsewhere for this opinion.[2] But, so far, physics still desperately leaves metaphysical problems to philosophers; and when physicists speak today of mind or consciousness, it is only to say something more about the structure of good old matter, and matter only, not about mind itself. However, this position is not so easy to maintain, as we shall see, when you accept that mind and matter should now be considered as two interacting "objects."

Let us return to Descartes again. It has probably not sufficiently been stressed that the French philosopher, when making his historical distinction between subject and object, had indeed given a dual structure to the object: object was made of the association of *thought* and *extension*. And the subject, the "I", was situated at the "intersection" of thought and extension; the "I", the subject, was not thought, but the one able "to look" at the object, that is look at both thought and extension. This is why "I" think, but also "I" know that I think; "I" know because "I" am distinct of thought, including my own thought, and consequently "I" may contemplate that thought, in the same way as "I" may contemplate extension (space-time):

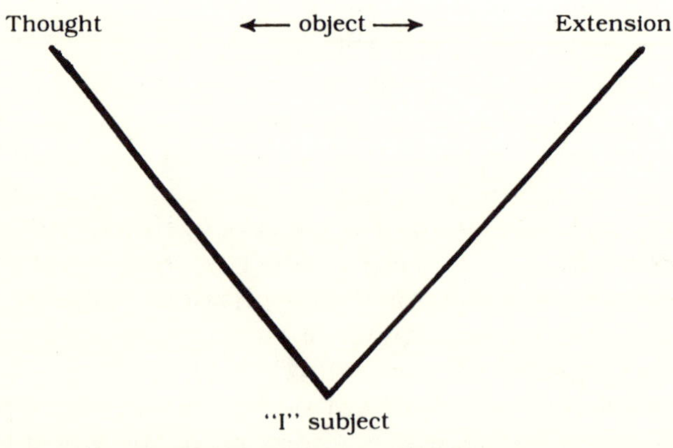

Figure 3–1: Thought-Object-Extension-"I" subject.

Without doubt, thought is not directly "observable," or at least observable as an ordinary object; let us recall the famous words of Paul Valery: "We could poke about for centuries inside a brain and never meet a state of consciousness." But, even if not "observable," thought is "representable," which means it could suggest to the "I" a way to come to existence as an imaginary "model": for instance, a state of "anger" could be translated through an audio-visual model by a painter or a musician, and also through a poem (Valery should have known that!). And, indeed, physicists have recently been learning that they also are able to represent thought, because thought is primarily an "object," although an object different from the extension. Thought is not "observable" in space-time, but "representable" through an "imaginary" model, a model suited to describe representable, although not observable, objects. The true "revolution" that is occurring in physics today is the introduction of a new language, the language for the representation of the non-observable Imaginary objects, indeed the language of thought.

Physics and The Existence of Imaginary Objects

Obviously, we already have found an existing imaginary object: thought itself. Who would seriously doubt that thought "exists"? Physicists themselves do not, but they have so far refused to accept thought as an "object." They probably would not have missed that important point so long if they had read Descartes more carefully. This ignorance was, however, a good excuse for physics to designate thought as being outside the field of investigation of an "exact" science, which should only deal with observable and "concrete" objects. Especially since the beginning of this century we have been attending to the development of a physics exclusively preoccupied by "observable" phenomena, an approach culminating in quantum mechanics.

Far be it for me to refuse to acknowledge the immense progress that quantum mechanics brought to the representation of phenomena, especially by eliminating some "false images" that were sustaining the basic axioms of what was called "classical" physics. But, as Einstein so long maintained, the probabilistic representation of phenomena was a "reductionistic" representation, in the sense that it was not a *complete* representation of our universe. This began to become clear to physicists when they discovered that some objects of the world (besides thought!) definitely "exist," but are, however, non-observable objects. Such is the case, at the cosmic level, with "black holes," which represent the final state of evolution of a great number of stars. These stars become invisible (that is no longer "observable") at the same time they go on "the other side" of our ordinary space-time. Such is again the case, at the microscopic, elementary particle level, of quarks, which enter the constitution of nucleons composing the nuclei of any atom of our world. The quarks "do exist," but will never be directly "observable." What should be understood clearly here is that these objects, black holes or quarks, are not "non-observable" because our scientific technology has not yet reached a sufficient level; they are non-observable "in principle," and the justification of their existence "includes" the reasons why they would never be "observable." These objects can truly be qualified as imaginary objects, that is objects which are "representable," but not "observable," that is defined as an extended region of ordinary space-time, the space-time of reality. Such objects can only be thought; indeed they are thought.

Like Descartes, present day physics confirms that there exist two types of objects: thought objects and extended objects. Consequently, we are led to redefine the field of investigation of contemporary physics:

Universe = Reality + Imaginary

= Extension + Thought

We shall agree to call reality (extension) the "observable" phenomena, and imaginary (thought) the "non-observable" phenomena, meaning here non-observable in principle. Both reality and imaginary are "representable," but reality is representable as consisting of extended objects of our ordinary space-time (the visible world around us), while imaginary is definitely not representable as an extended region of this ordinary reality. It is the main achievement of complex relativity to have extended Einstein's general relativity to existing non-local space-times, i.e. space-times not included within the reality space-time. In that extension of Einstein's space imaginary objects such as black holes and quarks . . . and evidently also thought, can now be represented.

Imaginary is Multifold, Reality is One

We are not here to enter into the details of complex relativity: this theory has now been published[3, 4] and specialists can therefore directly refer to these publications. However, we would like to mention here its main features relative to the representation of imaginary.

One of the principal characteristics is the multiplicity of imaginary and the oneness of reality. We mean that reality is made of one continuous universe, made of the cosmological space-time observable around us, containing all the galaxies, the stars, the planets, and all things living on these planets. None of these "things" belonging to reality can "get out" of the space in which they are immersed; reality is a "closed" space-time. If you decide to try to escape, flying "straight ahead" for billions of years, you would end up returning exactly where you started, as if you had moved "straight ahead" on the curved surface of a sphere. We say, with astrophysicists, that our cosmological, observable reality has a curved and closed space, so that there are no means for things contained in this reality space to be truly "separated."

According to Einstein's general relativity, all objects consist of one and only one "substance," space-time. Even what we call "empty space" still consists of space-time; thus all the objects you see around you in the vast cosmological reality, although they may sometimes appear to be separated, are indeed inseparable one from another. Consider the analogy of two icebergs: they are never truly "separated," for they are made of frozen water and it is the same substance, water, that comes between them. This notion of "non-separability" between all the objects of reality, whatever their distance, is a fundamental aspect of contemporary physics.[5]

We can say that reality is a collective and unique space-time, the objects extended in reality being different "shapes" produced out of this unique "substance": space-time.

On the contrary, imaginary is multifold. Each imaginary object is a small "personal" universe by itself, with its own space and time. If we think of reality as a large balloon, then imaginary objects would be analogous to bubbles collecting on the outside surface of the reality-balloon. They are "personal," in the sense that none of the bubbles shares space with any other bubble or even with the reality-balloon. We shall better understand how imaginary objects each make up a "personal" micro-universe when we discover that imaginary objects have "imagination"; and imagination is definitely a "personal" quality that is never shared with anything else.

An Imaginary Able "To Imagine"

Probably the most important result of complex relativity is to have shown that the imaginary "bubbles" actually have "imaginative" properties. At the microscopic level the imaginary objects all have the same structure and are called eons.[6] The model of the eon looks very much like the model of the cosmological reality, except that it is much smaller, "pulsing" much faster and much warmer. Eons are pulsing micro-universes closed on themselves; thus nothing can emerge from the eonic universe, exactly as it is for our cosmological reality. The "radius" of the eon is of the order of 10^{-13} millimeter which can be compared with the few billion light-years' radius of the cosmological reality. Similar to reality, the eonic universe pulsates, but billions of times faster. And again, like cosmological reality, the eonic universe is full of "light," or rather full of this special light that astrophysicists call "black-body radiation," but the temperature of the eonic light is billions of times warmer than the present cosmological temperature (2.7 Kelvin degrees). Finally, each eon has a space-time of its own, but the eonic space-times have properties that are complementary to the properties of the reality space-time, since the times of the eonic universes are "negative" times. The fundamental consequence is that in each eonic universe the phenomena spontaneously evolve to increase their "order" (decreasing entropy), instead of spontaneously evolving towards more disorder (increasing entropy), a well-known characteristic of our cosmological reality. It is this last characteristic which confers to imaginary objects, that is to eons, their essential new property: the photons of light that fill up the eonic universes are able to "memorize" and spontaneously increase the order (i.e., the meaning) of the memorized data by integrating these data (as we do when integrating words into a sentence). In brief, the imaginary objects called eons are capable of *memorizing* and *reasoning*. Exactly like you and me!

A Head and A Body For Each Particle of Matter

Now, where are those strange imaginary objects called eons actually located? Where can we practically meet them in our own universe?

This question is similar to the query: where do we meet consciousness in our universe? As we already have pointed out, consciousness is not directly "observable;" however, it is undoubtedly there, somewhere, in our universe. Complex relativity, by giving physics a way to represent the existing non-observable objects, has for the first time allowed a possible way to situate, one versus the other, the observable and the non-observable; matter and consciousness; reality and imaginary.

Complex relativity shows, and it is one of its important results, that there does not exist in our universe what we previously called particles of "inert" matter. Any particle, whether stable or unstable, does possess what we may call a "body" (although eventually point-like) in reality; but this body is always associated with a non-local, non-observable "head," made of one (and sometimes many) imaginary eons. To simplify, we might say of matter what biologists have for a long time been telling us concerning living organisms, from ants to human beings: they are made of a body associated to a conscious head. This unique privilege of living organisms is found at the simplest level of organization, the level of the individual particles of matter. Life is therefore the privilege of everything in our huge universe, which is a living being in all of its parts. Nature, which largely demonstrates the perfection of what it is able to invent, has decided to start with perfection from the simplest structures, so that the largest, which are made of the simplest, will benefit from the same perfection. (Upon reflection, "living" particles of matter would not surprise us so much if we could only rid ourselves of some strongly preconceived ideas. But as Aristotle remarked: "When faced with simplicity Men behave like bats facing light: they become blind.")

Particles of matter are of two fundamental types. Some, such as electrons, have just a point-like extension in reality, and are consequently and practically pure imaginary, i.e., have just an eonic "head." Other particles, like the nucleons entering the nucleus of the atoms, have a truly extended structure in reality, but are also associated with one (or many) eonic "heads," called quarks in contemporary physics. Both these particles are well-known to physicists today . . . but with the heads missing! The non-observable imaginary heads have so far lain forgotten in the drawer of the metaphysicists.

Verification of the Existence of Imaginary Objects

How can we proceed to demonstrate the actual existence of imaginary objects like eons? Well, we must show that the addition of imaginary space-times to the present landscape of physics allows us to complete our present knowledge of the universe. Physics has so far opened a window toward reality. Let us show that there is another window, and that it opens on another part of our universe—imaginary. If this can be done, if imaginary can be demonstrated as an existing part of our universe, then we should at the same time accept the new properties associated with imaginary and consequently also accept the principle of elementary matter linked with memorizing and reasoning properties.

To complete in this way our knowledge of the universe, we must first examine what is clearly missing from this knowledge. Lo and behold, physics today is still organizing itself around a large "blank." The representation of the phenomena can only correctly fit the observational data if we introduce into what we call the laws of nature a certain number of numerical parameters which, so to speak, reflect the actual experimental measurements which can be made of the universe. For example, a number of laws make use of a characteristic velocity, called velocity of light, which is approximately 300,000 kilometers per second. Why 300,000 rather than 30,000, or one million? No theoretical reason justifies the

300,000, and we are obliged to say that this number is "given" to us by Nature. This is another way of accepting that a Creator (here called Nature) has chosen by himself this very peculiar value of 300,000 kilometers per second for the velocity of light. We cannot help thinking that the Creator, at the time of Creation, certainly had better things to do than enunciate a series of numbers. Physicists are unenthusiastic as well about accepting a Creator who interferes with their "exact" science and would definitely prefer not to have to consider such an eventuality. And of course the velocity of light is not the sole number determined by the "arbitrariness" of the Creator. There is also the Planck constant, the masses of the particles constituting all matter, the electric charge of the electron—all in all, a good dozen "arbitrary" parameters (called the fundamental constants of physics).

One way to get rid of this embarrassing, "intervening," number-choosing Creator would be to suggest that while these numerical values are not yet justified by physics itself, it is only because physics is still lacking a part of the complete set of the laws of Nature. When we someday complete the set, then we will also find out that all laws are compatible, though *only* if the "fundamental constants" get the precise values that observation (and the Creator) had so far delivered. In the physicists' terminology, we would say that the complete set of the laws of Nature accepts, for its compatibility, a mathematically unique solution, which yields in turn the numerical values yesterday still considered as chosen by the "arbitrary will" of the Creator-Nature.[7]

This is precisely what happened when complex relativity completed the laws of the universe by adding the representation of imaginary to the representation of reality. This new window opened into the universe has revealed a large and still unknown region of our cosmos. And the compatibility that is required for the complete set of equations now representing the universe, with the old equations representing reality and the new equations representing imaginary, has yielded a *unique* solution that has discovered the numerical values of all the fundamental constants of physics, from the radius of the nucleon to the radius of the universe as a whole, including the velocity of light, the Planck constant, and the elementary electric charge of the particles. No more arbitrary numerical parameter enters the laws of Nature as described by complex relativity.

We could not hope for a better verification for the actual existence of imaginary. Yes, I know that it will still be hard for some "old school" physicists to grant a "head" to each elementary particle of matter and introduce a "conscious" dimension at the basic level of our universe. But we should sooner or later, reconcile ourselves to this idea, since it is today the more "complete" available representation of the world where we live.

Conclusion

Obviously, these results have important consequences, not only in physics, but also in biology and philosophy. The discovery of imaginary by physics can truly be compared to the discovery of a new world, indeed the mutation from a "material" to a "spiritual" world.

Notes

1. John A. Wheeler, C.W. Misner, K.S. Thorne, *Gravitation* (San Francisco: Freeman, 1973).
2. Jean E. Charon, *L'Esprit et la Relativite complexe* (Paris: Albin Michel, 1983).
3. Jean E. Charon, *L'Etre et le Verbe*, ed. du Rocher (Paris, 1983).
4. Jean E. Charon, *Complex Relativity and the Unification of all Four Physical Interactions*, in course of publication (Paragon House, New York, 1987).
5. Jean E. Charon, *L'Esprit et la Science, Colloque de Fes* (Paris: Albin Michel, 1983), *Spirit and Science* (New York: Paragon House, 1985).
6. The word *eon* was the name given by the Gnose of the first century of our era to the particle bearing "consciousness."
7. Geoffrey Chew, physicist at Berkeley University, California, has explicitly used the same argument to "measure" the completeness of a theory. Is a more "complete" representation of the universe a theory that "fits" the universe as well but makes use of less arbitrary numerical parameters.

Discussion

Jean E. Charon

Elizabeth Rauscher: I sympathize with your ideas which are very stimulating, and as you know I myself am working on complex geometries. I have one or two specific questions which I would like to ask you, besides many reflections which your paper suggested to me.

One conclusion of complex relativity, as I understand it, is that each particle of matter, be it electron or nucleon, whatever substance this particle belongs to, possesses an element of conscience, a mental element. . . .

Jean Charon: Yes, that's it exactly. Complex relativity leads to a representation of a particle of matter that includes not only what it was considered to be before, that is, a part of matter, strictly speaking, but also a part that represents the mental structure of the particle. It is a little as if we described at the same time in this particle an "outside," which is the old representation of matter as we have known it, and an "inside," which had never been described until now, and which complex relativity allows us to represent. But each particle, in its mental aspect, differs from every other particle, in the sense that the content of this mental part is not the same in one particle as it is in another, exactly as what is in your head and in your memory is not identical to what is in your neighbor's, or mine. We can thus say that there are particles which possess only a few elements of memory, a little like a newborn baby, and others which possess much more, depending on the lived experience of the particle. Of course, all of this, which I am speaking of in slightly pictorial terms, and by comparing it to what we, as beings of body and mind, possess and feel, all this in complex relativity is described in a precise way and represented in precise mathematical equations.

Elizabeth Rauscher: I think that this idea is natural, and I have felt it intuitively for a long time in my own research. An interesting concept, and one where there is often confusion, among the physicists themselves in particular, is a concept which posits that the symbols that represent our exterior universe are not those which make up this exterior universe, but are in fact symbols which exist *in our mental consciousness*. This is a distinction that it is important to see clearly. They do not claim that the exterior universe is made up of symbols, but that in order to understand this universe, we organize these symbols in our mind.

Jean Charon: Yes, that is important, and I am in complete agreement with you. Moreover, it is precisely that which makes any awareness that we have of the world purely relative, because this consciousness is made up for each one of us by the way we put together our own symbols of our awareness of the universe. Your symbols are not my symbols; the theologian will not represent the universe the same way a physicist does. We must be aware that the representation that each one of us has of the universe is completely relative, and relative to each one of us.

Elizabeth Rauscher: And these representations are not mutually exclusive but complementary.

Jean Charon: Exactly.

Elizabeth Rauscher: I appreciate the interesting approach you have to the representation of the world by physics, to which you add, notably, Godel's theorem, showing that the progression of physics must necessarily lead to a paradox, but that this paradox does not call into question the progression of physics. It simply indicates that we have reached a limit, and that from this paradox is going to spring a new progress, precisely by overcoming this paradox.

Jean Charon: I agree completely, and we must liken this to Diane Cousineau's paper on values. Really, there are no absolute values; there is continuous transformation, and these transformations lead from an established order to a complexity which itself results in a disorder. That is the paradox, but from this disorder will afterwards be born a new order. That is the same progression you have in physics: you advance from something you think is well ordered, and then the analysis and the research lead to such a great complexity that at a certain moment the disorder, the paradox appears. From that paradox, when physics continues to transform itself, you are going toward a new order, which is one more step forward toward an increasingly better knowledge of our world, I would even say, from the perspective of contemporary physics, that it is knowledge not only of your exterior world, but also of your interior world because, just as important, the mental faculty is now an integral part of the field of investigation in physics.

Elizabeth Rauscher: Naturally I was extremely interested too, and this is a crucial point, by the contribution of complex relativity to the calculation of all the physical constants in nature. This is really crucial. This is truly the value of your theory. You show that, first of all, all these constants in nature are linked to each other, which is what physics should have expected to find, but which had to be demonstrated, and you use your theory to calculate all of these constants. That is unquestionably a great mathematical contribution, and, in fact, an "experimental" proof of the model that you have developed.

Jean Charon: Yes, these constants, moreover, if you want to justify them by their "generation," if I can call it that, represent a true collective memory of particles, that is, that each particle has in its memory the sum of its constants. And I ally myself here with the recent ideas of the physicist John Wheeler, who thinks that the constants that are part of the laws of nature have been progressively born out of an original chaos where matter was interacting at the beginning of the world, at the time of the "Big Bang," without any laws present at that time. The laws and the constants appeared little by little in the course of a kind of "dialogue" among all the particles of matter present in the universe.

A member of the audience: I was interested in the ideas that you presented. However, I have a problem with the archetypes. In what way are they there, since you say that *being* is made up of archetypes?

Jean Charon: Your question comes back to the attempt to give an exact definition of an archetype because to know how archetypes came into a being, you must know what an archetype is. The notion of archetype, a term that the psychologist Carl Jung used very much, has a definition that is not easy to grasp. I think that, to get a better idea of what it is, we should try to see how the archetype is related to symbol and language. I will take an example, the example of a symbol represented by a cross, that is, two lines which intersect each other at a right angle. If this symbol is joined to a convention of language, it appears as an element of a language. For example, 2+2 = 4: in this example, all over the world, one will tell you that the sign "cross" means "plus," and one will give the definition of the word "plus" very clearly. Now take this cross in a way that I, personally, could use it as a mark on a page of a book in a certain place; I alone know the meaning of this cross. Why did I put this cross next to a certain word or a certain sentence in this book? There, the cross is truly a symbol, a personal symbol, a symbol for me alone. This symbol now gets its meaning not in relation to a collective language, but in relation to the very elements of my own conscience, and more specifically from the elements of my own memory. In other words, in my memory I have a number of interconnected symbols, linked to each other, and as for the symbol of the cross I just mentioned, I give it its meaning by the relationships I establish between this symbol of the cross and the other symbols in my memory. From the collective symbol, which is generally called "language," because there is, underlying the symbol, a convention among readers, a convention of language, we have arrived at a personal symbol which only concerns me.

But let's go even further back in the essence of things. Let us ask ourselves what happens to this cross if I omit any relation between it and the other symbols in my memory. Then, the cross has no more meaning. It will float in the air without being connected to any other symbol; it will lose its meaning.

But have I really omitted all the other symbols in my memory? Among the symbols in my memory, there are those related to space and time, because space and time are also symbols. Let us omit them. What happens to the cross? The cross does not exist, either in time or space. It almost disappears; it becomes nothingness. But this nothingness, however, is not no thing, because if I reverse the sense of time in the operation I have just performed, I must admit that there remains some thing which, as soon as I put back the symbols of space and time, will become a sign, but without meaning as yet; then, afterwards, when I put back in the symbols of my mind, this sign will become a cross, with meaning for me; and finally, when I insert this cross into a language, the language of mathematics, it will become the sign +. That is a collective symbol. Well then, what it remains when going back to the essence of things, this nothing which possesses the potential property of being able to "become" something by means of symbolic elements, that is, by means of the word, I would say that this left over bit which has only the possibility of "becoming" is the archetype. And, as we have just seen the archetype is this nothing, and that is why the archetype is so difficult to define. It only has for itself the property *of becoming*, and this should not surprise us when we try to define the archetype, because we are in the position of the Creator at the moment of creating the world: it is the transformation of the nothing into the Word, the transformation of the darkness into light. We perceive that the nothingness is necessary to create a world, that archetypes are necessary to make symbols, to make the Word, to make the universe.

José Delgado: We are composed of a multidisciplinary group here, and I am afraid that we may each remain in the straight jacket of our particular discipline. It seems to me that we are each speaking in our own language, and that this language is difficult for others to understand. I am not sure whether we are making a great enough effort to communicate, which we should try to do, because we are discussing certain fundamental things, like the brain, our mentality, our mind, but we are speaking of all this in our specialized disciplines like mathematics, physics, psychology, neurobiology. I wonder if it would not be useful for several of us to sit down at a table one day, and to try to see how it would be possible, in a general way, to communicate to each other the languages of these different disciplines. And I would personally like to be specific about my own difficulty in trying to understand everything well.

You spoke, and this is one of your essential ideas, about psychomatter. As a psychologist, I do not understand exactly what you mean by that. I agree that it is my problem and not yours, but it is really a problem for me. As a neurobiologist, I learned that matter had a certain number of properties, like mass and weight. The information which concerns the psychological and the mental does not possess any of these qualities. Mental means knowledge, establishment of relationships, and many other things, but never mass or weight. Then how can I speak of particles of psychomatter? I am not saying that you are wrong at this point. I would just like to understand why I am probably wrong in asking this question. And I am saying all this to show that we must absolutely establish a clear line of communication between us. If we do not, then a conference like this one will have only a limited interest. We are succeeding in speaking in French and English here, and in spite of these two languages we are succeeding in understanding each other, because there is an effort being made at translation. In the same way, it seems to me indispensible to have this type of communication among the various disciplines, and especially when we want to touch on these fundamental problems.

Jean Charon: I understand this difficulty in communicating among the disciplines. I am completely aware of it, and I think, as you do, that it is really a fundamental problem. But as for the specific problem of psychomatter, I think that we can alleviate the difficulty in communication with relative ease, by omitting many of the essential preconceptions that we have about matter. I think I can explain with an image. Let us suppose that a thinking being can come from another galaxy and approach the Earth, and come within several hundred kilometers of the surface of the Earth, and his instruments allow him to see the surface of the Earth and its objects, which are people, who are moving around on the surface. He will begin by pinpointing and analyzing whatever is visible, that is, the body of these humans that he sees at the end of his telescope.

And, if he has enough knowledge, he will think that these humans are made of matter, a matter that is susceptible, as you yourself propose, José, of characterizing itself by physical elements such as mass, weight, energy, and many other things. But these are only the results of a very first step toward knowledge. After a little while our visitor from another world will say to himself that these objects of matter which are people must possess something which cannot be seen immediately, but something which they use to communicate among themselves and to accomplish a certain number of tasks. That something is thought. Then, of course, if he is even a little bit of a biologist, he will end up discovering that this non-observable thing called thought is emitted, so to speak, from a certain part of this body of matter of the person, and that part is called the brain. Our visitor will then understand that these objects that he has been observing from his interspatial machine are composed, on the one hand, of a body of matter where the classic laws of matter are applicable, and especially the laws governing mass, weight and energy, and on the other hand [they are made up] of thought, that is, information, which is not subject to the characteristics which apply to matter. Here he must speak in terms of symbols, of communication of symbols, of memory, etc.

Well, contemporary physics is at this same point as our extraterrestrial visitor, who has made the analysis I have just described about humans, who are endowed with the two great but different characteristics, which can be called the material and the mental, or, if you prefer, the real and the imaginary. Complex relativity does not discover any new matter or any new particles; it is still the same old matter that it is looking at. So why do we call it psychomatter? We call it psychomatter because, as in the example we have just given, we are discovering that this matter has characteristics which are not solely material, but also has characteristics of another nature which are mental characteristics. These mental characteristics, like thought in humans, are not able to be shown immediately in terms of matter. We need special mathematical tools to make them appear. Roughly speaking, we will need tools which will allow us to grasp the real *and* the imaginary at the same time. And this is precisely the formalism of complex relativity. In other words, even though, like our visitor from another world, physics has for a long time and even up until now been content to deal with what could be observed, and even almost legislated that it would only deal in its representations with what was observable, this same physics is discovering suddenly that it is necessary, in order to give a more *complete* representation of matter, to establish a frame of reference which will allow the description of the observable and that which is not directly observable, but which can be represented. I say not directly observable and not simply unobservable, because if thought were totally unobservable, well, why would we speak about it? Thought can be represented but is not directly observable. It is the same for the entire imaginary part of representation that complex relativity offers about matter. Thus psychomatter is still the old matter. To use a quotation from Teilhard de Chardin, who has been shown to be a precursor in the physics research which only appeared twenty years after his death, I would say that in matter we have been describing until now only the "outside" of the particles of matter, the observable was represented, and that all of a sudden we discover that, in order to be more complete, there is also an "inside" which we must also represent, even if this "inside" is not directly observable.

That, my dear José, is what psychomatter is: it is still your good old matter of biology, but the physicist's eye is now giving it a more complete description. This does not take anything away from neurobiological studies, of course. I think that biology should welcome, I would even say, "with open arms," the fact that physics can now say something about mental characteristics at the level of each particle of matter.

One last remark: I do not think that anyone would dare suggest such a "thinking" model for particles if he did not have, as support for this proposal, *extremely* sound proof. That is why I myself have insisted on proofs for complex relativity which are based on calculations, by means of equations of this model of a particle of psychomatter, the calculation of the *totality* of what we call the fundamental constants of physics. As Elizabeth Rauscher has just noted, that is naturally the crucial element in the demonstration. And that is really the single argument which allows us to consider so seriously the prospect of psychomatter.

Mitsuo Ishikawa: You verified the model of psychomatter by using equations on this model, but what hypotheses did you use to construct these equations?

Jean Charon: Here is not the place, of course, to enter into the mathematical details of complex relativity, and I must ask you to look at the works I have published on this subject. But your question is important. Complex relativity borrows (as Geoffrey Chew did in his theory of the Bootstrap) from principles of *global coherence.* That was also, as you know, the basis of the axiomatic of general relativity. These principles of global coherence are, for instance, principles of conservation, affirming "nothing is created, nothing is lost"; on axiom-definitions such as "what follows must come after what goes before"; or: "a phenomenon must be such that all its possible states must have a total probability equal to one"; as well as several principles of homogeneity and isotropy. In other words, nothing could be supposed which would not come from the simple consequences of the *logic* that we use to write the formalism of physics. I think that complex relativity is the perfect illustration of what Einstein wrote about in his autobiographical notes at the end of his life: "I have learned something else about the theory of gravity: no totality of experimental facts can ever lead to the writing of equations as complex as those of relativity. *A theory can be verified by experimental data, but there is no way that leads from experimental data to the creation of a theory.*

Complex relativity, which is based only on the principles of global coherence, seems to me, as I said, a perfect illustration of Einstein's affirmation. And it is here no doubt that modern biologists and physicists are the most divided, because most recent biologists in practice and by nature always go from the experiment to the theory. For the contemporary physicist, the experiment comes in second place, and first comes the logic used to write anything, and notably to develop a "model" of the phenomena of our universe. Logic and only logic! It is true that you can change your logic, but in that case you will also change your universe. I think that is what Ashok Gangadean told us in his wonderful paper on grammatology.

Ashok Gangadean: I am truly aware of and amazed by the fundamental change you have brought about in the approach made by physics to the universe. However, from the ontological point of view, I think that you have done a classic piece of work, which no doubt needed to be done by contemporary physicists. I note in your work very significant echoes of the approaches of Descartes, Spinoza, Leibniz and Hume, who also tried to offer a coherent representation of the world, and, notably, of the existence of the material and the mental. I also appreciated the extremely important semantic definitions that you gave at the beginning of your paper, which entail a revision of the modern language of physics. The most surprising juxtaposition with the "ancients" is no doubt with Leibniz's theory of monads. These monads were also microcosms containing light, which formed the basis of the mental process itself.

One troubling thing, however, which is a problem for me, because this is the problem of the whole post-Cartesian epoch, is to know if you, for your part, have obtained a unitary or a unifying description of representation, or if you were led to consider, as Leibniz did, *two* parallel universes, two parallel languages of representation, which are the representation of matter and the representation of the mental, finally, the representation of the real and the representation of the imaginary. These implied for Leibniz a sort of "preset harmony," so that what happened in one universe had immediate repercussions in the other. My question is thus the following: do you have in your case two universes, with a pre-existent harmony between the two, or just one universe? In other words, is your psychomatter astride two parallel universes, or is psychomatter really a fundamental element unifying representation?

Jean Charon: Complex relativity is a unified theory, in the sense that it describes *one single* universe. Moreover, its major result from the standpoint of physics is, as I have just insisted, the calculation of the fundamental physical constants in the real physical universe, in the one and only universe which is ours. However, what complex relativity shows is that this universe carries in it the collective and the personal at the same time. The collective is the old observable representation that physics gave of the universe. That is what I have called reality, or which is still called the cosmological universe, that is, this immense universe filled with stars and planets which surround us and which we can perceive. But with complex relativity, this cosmological universe appears as a sort of place for a dialogue of the mental, this mental part being itself made up of micro-universes, that is, entities which are capable of collecting *personal* information. Let us note that this characteristic of personality was necessary if you wanted to give a real representation of the mental which corresponds to what we know about it. What you have in your head is personal to you; you do not share it directly with what is in your neighbor's head. Well, this trait is established at the level of every particle of matter: whatever it has "in its head" is particular to it alone, belongs to it alone, and does not mix with what the neighboring particle has in its mental set-up, in its memory. Yet, these two particles can have "exchanges" in the collective cosmological universe, that is, they can have a real dialogue between themselves, and in so doing can release personal information that they possess in the collective universe, and in this way exchange information in the course of a real dialogue.

I know that all this, which would seem perfectly natural if we were talking about humans, or even about basic living particles like bacteria, seems at first glance extremely surprising, because, at least since Newton, physicists have considered that particles of matter were comparable to simple "billiard balls." But today, it seems that we must attribute to them qualities which are both material and mental. This, it seems, can only make our task much easier, our task being to interpret the results of experiments according to which matter tends to *organize itself* spontaneously, as it also does in living things.

Ashok Gangadean: Do you think that the notion of psychomatter and complex relativity can help us understand people's reactions, those reactions called psychosomatic, that is the direct interaction of the mental and the material in the human body?

Jean Charon: That is precisely a task which should come as a special priority to the biologists. As a physicist, I would simply like to say what I found on the level of examining the structure of particles of matter, which are particles of psychomatter when represented not only externally but also internally, in their material and in their imaginary or mental part.

4.

The Interactive Mind in the Participatory Universe

Henryk Skolimowski

> *The universe does not exist "out there", independent of us. We are inescapably involved in bringing about that which appears to be happening. We are not only observers. We are participators. In some strange sense this is a* participatory universe.
>
> John Archibald Wheeler

> *Mind is indeed the source of bondage and is also the source of liberation.*
>
> Maitri Upanishad

Reconstructing the Background Knowledge Situation

The mind is one of the lasting mysteries of the universe. It is not the slayer of the real, as some Hindu traditions maintain. It is the creator of the real. Whatever we know, we know through the agency of the mind. The mystery of the mind is thus doubly profound: not only is it an extraordinary creation of nature—exquisite and puzzling in its own right; it is also the shaper and creator of reality. On the way the mind works depends the nature of our knowledge, and (a step further) the shape of the external world. What the mind cannot render, the world cannot bear. And what the world bears is exactly what the mind renders.

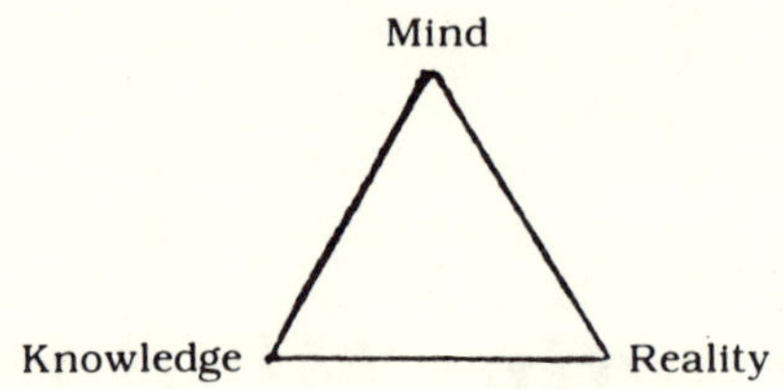

Figure 4–1: Mind-Knowledge-Reality.

It cannot be otherwise. The world is what our knowledge can describe. Our knowledge does not come *deus ex machina*. We find in our knowledge only that which our mind can conceive. We have been slow, and often obtuse, in perceiving the fundamental unity of mind, knowledge and reality. Many a time by separating the three elements, we falsified the whole picture. It should not escape our perception that a change of our outlook on one of the three elements necessarily implies a change on the other two. Admittedly it takes time before the adjustment is completed and the three find themselves in coherence and harmony. Let us now see how the interdependence of mind, knowledge and reality manifests itself within Western philosophy of the last fifty years. But first a brief historical interlude.

Parmenides knew it well: "No mind, no world." Immanuel Kant was equally astute when he postulated that the structure of our knowledge (and thus of our world) conforms to the categories of our mind, and not conversely, as classical empiricism had assumed. Kant's great insight was marred by one misconception. He made the structure of the mind too rigid and too dogmatic. Had he made the structure of the mind governed by some evolutionary principles, all would have been well. Instead he made it in the image of the inflexible categories of Newtonian physics. There is a paradox here, and an epistemological tragedy. Kant meant to liberate Newtonian physics from the straightjacket of empiricism. This he did. Yet in the process of intended liberation, Kant imprisoned himself in the rigidity of the conceptual universe of Newtonian mechanics.

The genius of Kant consisted in inventing the structure of the mind from which Newtonian physics *follows*. But this genius was also a part of Kant's undoing. When Newtonian mechanics started to collapse at the end of the 19th century, this spelled the doom of the absolute categories which Kant attributed to the mind. Already in the second part of the 19th century, non-Euclidean geometries were harbingers of bad tidings for Kant's categorial system.

One must be careful here. Advances in geometry (Lobatschevsky, Rieman and others), as well as in physics (Einstein and others) did not negate Kant's main thesis concerning the *active* role of the mind in shaping knowledge according to the mind's dispositions and capacities. What was undermined was the *specific structure* that Kant attributed to the mind.

Kant meant to finish empiricism once and for all. Yet, in spite of his marvellous insight into the nature of human mind, we witness the return to empiricism, wave after wave, as if Kant had never existed. If Mach's positivism was not enough, we evolved in the 1930s—under the auspices of logic and semantics—a new brand of empiricism, called logical empiricism.

The tools of logic and semantics have simply obsessed the minds of 20th century western philosophers. Through fastidious and exact systems of semantics, logical empiricists (and then analytical philosophers) have hoped to reconstruct traditional philosophy so that it would be on a par with precise systems of exact science. Surely, an impossible dream. But philosophy is full of impossible dreams; indeed it thrives on them. And so, having worked out elegant and precise semantic and logical matrices, philosophers in the 1930s decided that the structure of the world must conform to the structure of their propositions. Hence, *The Logical Structure of the World* (Carnap, 1928, 1962).

Now, let us clearly see the nature of their enterprise. The main problem for logical empiricists, for whom Bertrand Russell was the chief inspiration (particularly his *Principia Mathematica*) was how to secure the reliability of human knowledge. Their overall solution was *the structure*. In the right logical and semantic structures lies the key to the reconstruction of our knowledge; also the key to the reconstruction of physical reality. In this context, understanding means understanding of logical structures. The objective world is forced, *per fas et ne fas* into logical structures which our semantic systems provide. This is particularly striking in regard to Russell's/Wittgenstein's logical atomism, within the compass of which negative facts were invented to fit the structure of negative propositions. The epistemological and semantic theories worked out in the 1930s and 40s under the auspices of logical empiricism by and large conform to their norm: the logical structure determines the known, and the process of knowledge.

Within empiricist philosophies, the mind is curiously passive. The theories of mind stemming from logical empiricism, whether of Ryle or Armstrong, are impressive by their *incomprehension* of the mind. Seldom has such a sophisticated language shed that little light on the phenomenon it wished to describe. The reason is obvious. The materialist theories of the mind postulate that the mind is nothing but a kind of mirror in which the objective stuff of the world is, as it were, photographed. Such theories cannot succeed, as they make a caricature of the creative élan of the mind. Actually, the full story is more complex and more subtle. The creative élan of the mind is employed, within logical positivism, for the invention of logical structures. After the structures are invented and refined, the mind, in a sense, is deemed spurious. At this juncture the structure of the world must fit logical categories. And the structure of the mind must fit these categories. Thus the mind ends up being tortured on the procrustean bed of mental concepts that are logical categories invented for the descriptions of the activities of the mind.

The man who clearly saw the limitations of this excessively logical approach, who actually lived in the midst of the logical positivist revolution while it was happening in Vienna, was Karl Popper. He was one of the first to call the young Turks of the Schlick School the positivists. In his epochal book, *The Logic of Scientific Discovery* (1934, 1959), he showed the importance of the creative agency of the mind. Kant was a big inspiration for Popper, though Popper conceals it.

With Popper we witness another revolution in philosophy. To begin with, the main problem for Popper is to understand how knowledge grows. Let me reconstruct some of the premises of Popper's philosophy. To understand the world is to understand the nature of our knowledge. To understand our knowledge is to understand its growth and vicissitudes. Human knowledge does not grow like a pyramid, made of the same kind of stones (physical facts). The shape of the pyramid is continually reconstructed; as some parts of it are destroyed or at least significantly changed. The material of which the pyramid is made is also varied; all kinds of things are built into it.

For logical empiricists the chief inspiration was Russell and his *Principia*. For Popper the chief inspiration was Einstein and his theory of relativity. Einstein's theory has shown, according to Popper, that no knowledge is absolute. Even so an established theory like Newtonian mechanics finally marvellously gave way; it was not the repository of ultimate and unshakeable knowledge it had once been thought to be. The conclusion that follows is that all knowledge is tentative and conjectural. The best we can do is to play imaginatively with new conjectures (possible theories), and then submit them to relentless scrutiny. Our rationality, and indeed the foundation of our epistemology, lies in the process of relentless criticism rather than in the process of building an infallible rock bottom of knowledge out of atomic facts and propositions.

What the *structure* was for logical empiricism, the dialectical *process* of continuous criticism has become for Popper—the cornerstone of rationality and of epistemological worth. There is no question that Popper should be hailed as one of the intellectual giants of the 20th century, and a man who has helped us to liberate ourselves from the straightjacket of logical empiricism. Yet with the passage of time, Popper has become a new orthodoxy. His followers want in particular to monopolize truth and behave as if they had the license for all the right solutions.

What was especially liberating in Popper was his pluralism as he led us away from the one-track approach of logical structuralists. Popper has promoted pluralistic epistemology and open-ended rationality. It was the open-endedness of Popper's philosophy that was so refreshing and promising for the future. To try to make of Popper's opus a closed and untouchable system is to do violence to the very nature of his enterprise, according to which *everything* is open to question and criticism. It follows therefore that as our perspectives and problems change, so do our approaches and solutions. And our perspectives and problems do change. During the last 30 years Einstein's theories have ceased to occupy the center of our inquiries, either in physics or in epistemology. Instead quantum theory, or to use a more general term, The New Physics, has been posing most fascinating and mind-boggling problems both for physicists and epistemologists. Before we go into this realm, let us take a summary look at Popper.

In contradistinction to logical empiricism, Popper acknowledges the active role of the mind. The mind is constantly active, either by generating new conjectures or by ingeniously thinking up new tests of existing theories. But there are boundaries to this process. Reality seems to be given to Popper. At the center of his system is the notion of empirical refutability. You must have a firm and unequivocal nature (or reality) "out there" which will shout "No" to some of your conjectures. The idea that reality is given is reinforced by Popper's relentless defense of the classical (or the correspondence) theory of truth: truth consists of the correspondence between facts (phenomena) and their description. The underlying assumption of the correspondence theory of truth is that there is a firm reality out there which we grasp, or at least approximate in our theories. The match between description and reality is based on the notion that there is a "reality" prior to the description, which the description attempts to convey (capture, represent) by means of linguistic symbols.

Popper's philosophy was unrecognized for quite a while because of the dominance of logical empiricism. It was immensely helped by the appearance of Thomas Kuhn's book *The Structure of Scientific Revolutions* (1963), which is thoroughly Popperian in its approach. Popper's philosophy was further popularized by Paul Feyerabend and Imre Lakatos. These successive extensions of Popper's philosophy led to surprising, if not paradoxical, results. By the relentless criticism of the very basis of this philosophy, its practitioners have so much undermined the very notions of refutability and rationality that the whole epistemological enterprise has become problematic. The process of the historical reconstruction of science, within Popperian epistemology, has led to the undermining of every creed and contention that was held about science. Following the critical approach to its conclusion, Feyerabend has announced that *anything goes*. The enterprise of science, when examined penetratingly, according to Feyerabend, appears to be incoherent and amorphous and full of question-begging. Feyerabend ends up with "A Plea for the Hedonist," an article in which he argues that the world without science would be pleasanter to live in.

The watershed for Popperian philosophy came in the mid-1970s. The colorful and irrepressible Imre Lakatos may have been the man to bring Popper's philosophy to its eclipse, mainly because he did so much to popularize it. So intense was his zeal to put Popper on the map and diminish the influence of Kuhn, whom he saw as the major rival, that in undercutting Kuhn (see especially *Criticism and the Growth of Knowledge*, 1974 Lakatos ed.), Lakatos and others undermined the whole élan of conjecturalism. Today, whatever Lakatos' role in extending and therefore in diluting Popper's opus, Popper's philosophy appears to offer less and less for the newly emerging epistemological problems of quantum physics, especially for the conception of reality as co-extensive with the mind. Popper is an epistemological realist of the traditional kind, as he believes in a firm reality out there, a view the New Physics does not support.

We have recorded a famous confrontation between Popper and John Archibald Wheeler, in which—against Wheeler's claim that we live and think in a participatory universe—Popper is purported to have said: Your theory is contradicted by biology and by fact. Popper should have known better. He has always insisted that our facts are constructions of our theories. Given different sets of conjectures, different facts follow. And so it is with our present biological knowledge. It is by no means absolute or untouchable. As we reconstruct reality and the mind, we will need to postulate new conjectures, from which new facts may follow. This strategy is quite congruent with Popper's conjectural philosophy. There is more in Heaven and on earth than philosophers (so far) have dreamt of in their theories. We must not limit the fullness of our universe to the physical realm only. For if we do, we end up as reductionists and materialists.

Let us move to our times. We have travelled from the period dominated by the *structure* (logical empiricism), to one dominated by the conjectural *process*. We are now moving towards the period inspired by the notion of *imagination*, and the idea of the imaginary; imaginary dimensions and spaces are (according to Jean Charon) a part of the real and can be put into our equations describing reality.

Any new description of reality that attempts to reveal its new aspects is not going to be easy, particularly if it is to be carried out in the old language, that is to say in the language of the old paradigm. What, therefore, are our new problems? One of them is to make sense of subatomic particles too subtle to be described in *physical* terms. But this is only one problem of the whole cluster. Others are: How to redefine objectivity, so that, on the one hand, we acknowledge that we live and think in a participatory universe, and, on the other hand, we do not fall victims to the sick minds? How to make sense of the "knowing" electrons which, in double split experiments, seem to be aware of what some other electrons are doing? How to make sense of the new wholeness in which consciousness and the stuff this consciousness operates on are integral parts of each other? What is the meaning of wholeness within the new epistemology inspired by the New Physics? And what is the meaning of *parts* when parts are no longer considered independent of each other? How do imagination and reality interact with each other? What forms of imagination are translatable into reality? What is the limit to our imagination in conceiving new realities, and then in bringing them into being? Where is the dividing line between the imaginary, which may become a part of the real, and the fictitious which will forever reside in the recesses of our fancying? What is the nature of this form of imagination which can bring about new aspects of reality, and can this imagination be described in mathematical formulae, as Jean Charon suggests? What kind of language must we develop to make good sense for the imaginary which at the same time is (in one sense of the term) a part of the real? How far must we extend the meaning of the real, and thus the current notion of "reality"?

All these are new questions. They are outside the scope of either logical empiricists or Popperians. They outline a new epistemology. We must emphasize that as Einsteinian physics was a challenge to and an inspiration for Popper, so the New Physics is a challenge to and an inspiration for us. Einsteinian physics called for the creation of a new theory of knowledge, a theory that would recognize that no knowledge is absolute but would at the same time reconcile Newton with Einstein. Popper's conjecturalism was the answer. It proposed that Newton's theories were an approximation to and not an embodiment of absolute truth.

We are now at an altogether different juncture. What we need is not an update of Einstein but a different model of reality, and a new theory of the mind. This is then our program—to create a new model of the mind that truly makes sense of this participatory universe and our role in it. Needless to say, we want to create a theory both rational and coherent. Yet this rationality must not be at the mercy of the criteria of mechanistic rationality, which is a peculiar product of empiricism and a guardian of its conceptual universe.

From Tabula Rasa to the Interactive Mind

The new epistemology and a variety of theories of the interactive mind are being worked out simultaneously by a number of people, often without a clear awareness of the main epistemological problems which these theories attempt to meet. David Bohm's *Wholeness and the Implicate Order* is about the New Physics. But it is also an epistemological treatise, though not written within a conscious epistemological framework. The same can be said about the works of Capra, Charon, Brian Josephson, Prigogine and Wheeler. Each is preoccupied with his own problems. But their work is united by a common thread that spells out both the rejection of empiricism, on the one hand, (particularly its conception of the mind as tabula rasa), and, on the other hand, an acceptance of an indeterministic, open, participatory universe.

The thinkers mentioned above are physicists or ex-physicists. Yet the chief import of their work is not the extension of the physical but the creation of the metaphysical. We are in the midst of one of those exciting periods when working out a new outline of physics signifies working out a new metaphysics. Even those physicists who stubbornly cling to the old paradigm, and want to see in the New Physics "just the description of reality as it is," engage in metaphysical speculations on a huge scale. The present grand unified theory *postulates* the decay of protons. No such decay has been "observed." Indeed it has been demonstrated that within the scale of our universe (15 billion years or so), protons *do not seem* to decay. To save appearances (and the grand unified theory) physicists now postulate that although we do not have any evidence for the decay of protons *without our universe*, this should not worry us unduly. Why not assume that the period *needed* for the decay is 40 billion years or more? Thus the protons have been decaying before our universe came into being. To fancy that we can know what had been happening to protons before the birth of our universe is a metaphysical speculation in the primary sense of the term "metaphysical," and present physics is full of such speculations. We should not discard them as useless but acknowledge them as a necessary tool of our new groping.

The period of the empiricist innocence is over. And the time has come to work out a theory of the mind that not only recognizes its active role but also shows how it contributes to the creation of knowledge; how its creative nature makes it the same agency both in science and in the arts (surely, we do not have two different minds, one for science and one for the arts); how the early systems of knowledge and traditional cultures can be seen as products, extensions and manifestations of the same mind; and how different systems of knowledge can describe reality differently, although they are products of the same human mind. The remaining part of this paper spells out some characteristics of this mind (which can be described as holistic, participatory and co-creative), and the new epistemology that corresponds to it.

Concerning the Cosmos and the Mind

The outer walls of the cosmos are the inner walls of our mind. The dimensions of the universe correspond to the spiral of our understanding. There is no world given to us outside the categories of our understanding (comprehension). To know is to constitute the world. To apprehend, grasp, behold, seize the world is to embrace it in the tentacles of our knowledge. Beyond the tentacles of our modes of knowledge, the world is a buzzing confusion. . . . Actually, to describe the world as "a buzzing confusion" is to order it somewhat. To speak of the world is to commit it to our knowledge. To speak of the world is to sentence it to the categories of our knowledge. Reality is continually becoming; to say: "Reality is", is to commit a category mistake.

At those points at which the spiral of understanding (represented in red) pierces through the walls of the cosmos (as we

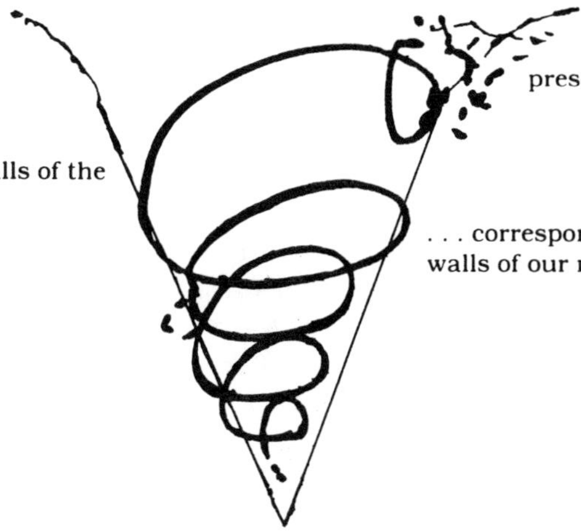

Figure 4–2: Those points at which the spiral of understanding pierces through the walls of the cosmos (as we envisage it), some extraordinary things happen: our understanding—at least the understanding of some individuals—has outgrown the confines of the cosmos as *defined* in our knowledge.

envisage it), some extraordinary things happen: our understanding—at least the understanding of some individuals—outgrows the confines of the cosmos as *defined* in our knowledge. We do not know what to do with those new insights—acts of comprehension that spill over the cone of the cosmos. These new insights are "beyond our world" until the walls of the cosmos are loosened and enlarged to incorporate them.

At present we find ourselves in such a situation. The new insights (of the New Physics and other disciplines representing the new paradigm) signify those parts of the spiral of understanding that visibly go beyond the walls of the cosmos as established by Western science. We do not know what to do with these insights—forms of understanding that are spilling over the accepted cosmos—until we enlarge our cosmos and find a new coherence. Then we see that these new insights are perfectly compatible with the enlarged cosmos.

Cultural anthropologists are aware that the cosmologies of various peoples differ. This means that their views of reality differ; also, that they use different kinds of knowledge to describe different realities. Sometimes we can translate their world views and their knowledge into our world view and our knowledge. Sometimes we cannot. In the latter cases we witness the *incommensurability* of realities and of forms of knowledge. Human history is full of examples of one people not comprehending the views of other people. Instead of condemning those whom we do not understand "for not knowing any better," we can now say—from the position of the new epistemology—that true understanding means compassion, means understanding that there are different roads leading to reality and to enlightenment. Older spiritual traditions were quite right to insist that compassion is the ability to see reality through the eyes (and hearts) of others. The new epistemology can now rationally justify this insight. The enlightened mind can accommodate the spiral of understanding of the other within his/her own spiral and can reconstruct the cone of the world view of the other; unenlightened and bigoted minds are locked in their own spiral of understanding. This also applies to cultures. An enlightened culture can imaginatively reconstruct the premises (the spiral of understanding) of other cultures and can demonstrate that ways and paths which at first appear "odd" and unacceptable from the viewpoint of our standards are perfectly normal and acceptable within those other cultures. Thus one of the consequences of the new epistemology is *tolerance*, which is a constituent of compassion; on another level tolerance is a precondition of imagination, which is a *sine quo non* of creativity.

Concerning Simplicity

The questions: what is simple? what is complex? are both exquisitely simple and exquisitely complex. The cell is simple when we consider it as a building block of an organ. But if we look into the cell itself, it is amazingly complex; it mirrors the universe. How do we comprehend the universe of the cell? By descending to its simpler parts, first to its molecular structure, then to its atoms. As building blocks of molecular structures of living cells, atoms are simple. But when we look into the structure of the atom itself, it too becomes amazingly complex. And the more we know about it, the more complex it becomes. This way of expressing the matter is perplexing, to say the least. Was the atom simple before we *made* it complex in the pursuit of our knowledge? Was the cell simple, before we *made* it complex by knowing more and more about it?

Now, the attributes "complex" and "simple" are not objective attributes of the world as it is. Why should the world bother whether it is simple or complex? In short, the patterns and configurations of the world are not there; they are not there independently of us. They are the patterns of our knowledge and the configurations through which our minds work. Things do not become either simple or complex—whether they are atoms, cells, or galaxies—by themselves. It is our knowledge that makes them so. It is our mind that makes them so. Some maintain (Eugene Wigner, for example) that our knowledge of physics has grown so phenomenally that no single individual can embrace it. They also maintain that no single individual can embrace all knowledge of chemistry. This augurs badly for our understanding of the physical world. Does it mean that the physical universe is so complex that it is beyond our comprehension? It may be so. Or does it perhaps mean that the *picture* we have created is beyond our grasp? I believe the latter is the case. We have made the picture of the physical world unmanageable and in many ways incomprehensible.

The mind cannot help imposing patterns of simplicity on the world, in order to understand the world. Such is the nature of our mind. Simplification is its *modus operandi*. Expressed in other terms: *Simplicity is the methodology of the mind*. Insofar as mind interlocks with reality and attempts to understand this reality, complexity/simplicity is a dual aspect of the world/understanding, or reality/mind. We do not yet have the right categories to articulate this insight. The present world, as rendered by categories of classical science, has grown so "complex" that it is beyond our comprehension; which means it is too complex for the mind. And that in turn means the mind will have to devise new patterns of understanding, new patterns of simplicity, new forms of logos, and impose them on this bewildering complexity in order to arrive at a new order and a new clarity. This is what mind has always done in history.

Concerning the Simplicity/Comprehension Thesis

Let me now introduce a new element of the epistemology of becoming (which is another name for the new epistemology), which I call the simplicity/comprehension thesis. With the emergence of the self-conscious mind, which is creating knowledge as a vehicle of understanding, the story of complexity/consciousness acquires a new human meaning, for it is continually punctuated by the stages of simplicity/comprehension. The human mind cannot cope with too many and too intricate complexities, and for this reason continually imposes its patterns of simplified understanding. If the cosmos is infinitely complex (and there is no reason to believe otherwise), if a single cell is infinitely complex (and there is no reason to believe otherwise), the mind cannot deal with them. Mind, therefore, imposes patterns of simplicity that constitute human comprehension. This process leads to what I call the metamorphosis of reality.

If reality becomes increasingly complex as the result of evolution's propensity to create ever more complex forms as its *modus operandi* (Teilhard's thesis of complexity/consciousness), then at some point this complexity will become so enormous that it will be beyond our grasp. I am here discussing the ontological complexity of the outside world, as if it were independent of our mind, see Figure 3.

But the ontological complexity is not independent of our mind. In actuality we witness two parallel orders: the ontological order (the order of reality) and the epistemological order (the order of the mind). Complexity on the ontological level and simplicity on the level of our understanding (on the level of the mind) are inseparable companions. We always simplify in order to understand. Understanding *is* simplifying, and human knowledge represents the patterns of simplicity through which the richness of the universe is digested by the human mind.

When the complexity of the ontological order becomes unmanageable, the mind "simplifies" reality by imposing on it a new order. A new form of logos is imposed on the unruly cosmos. This is, for example, what happened around the 6th century B.C. in ancient Greece when the mythological conception of reality gave way to a rational conception of reality. This was not a unique event in the history of Western thought but a

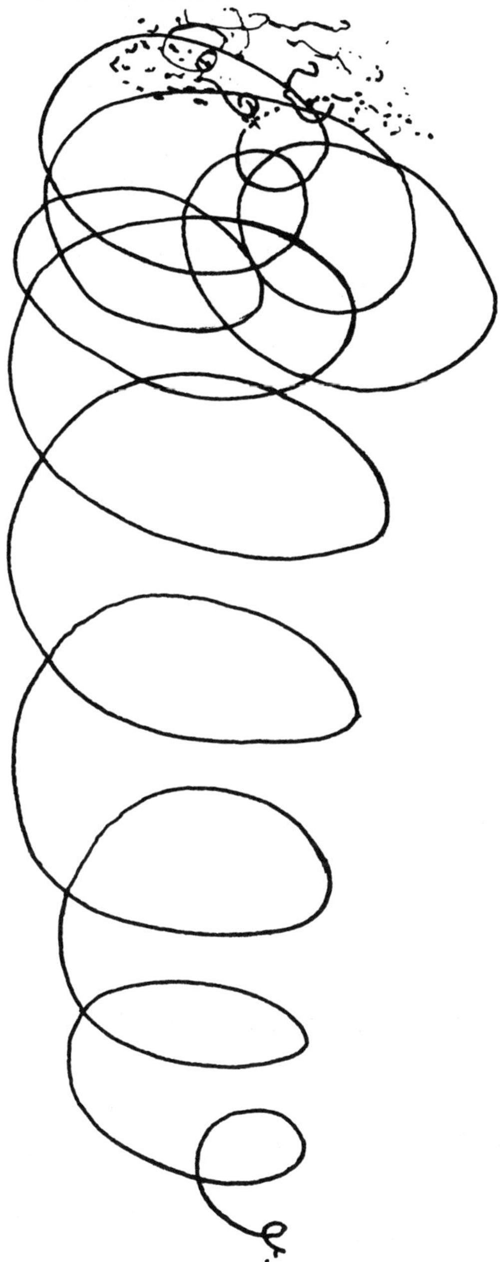

Figure 4–3: Ontological complexity of the outside world, as if it were independent of our mind.

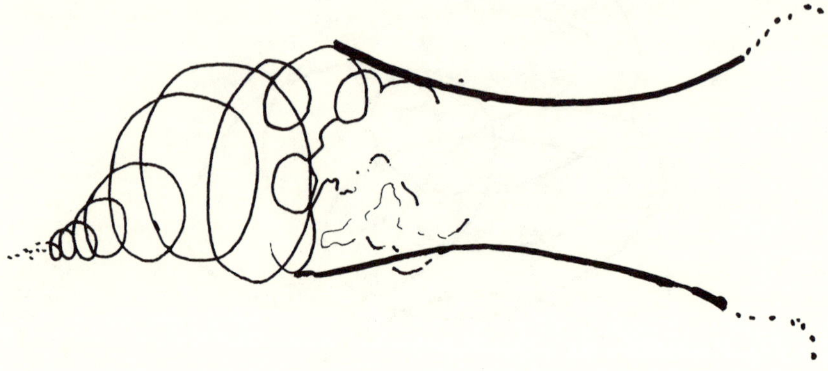

Figure 4–4: When our knowledge becomes too complex, the mind imposes a new logos on past complexities.

part of the pattern which was to recur again and again. The important point is that the order of the mind continuously shadows the order of reality. In the final analysis the two interlock with each other inseparably. At those periods when our knowledge becomes too complex, the mind imposes a new logos on past complexities. See figure 4.

Let us examine some of the important junctures in the history of Western civilization, in which the mind has imposed on reality a distinctive pattern of simplicity. We will look at four basic cycles and corresponding forms of logos (simplicity) during the last twenty-six centuries of Western culture, and will see how the various forms of our culture existed in the image of the predominant logos that shaped their characteristics.

The Graeco-Roman cycle begins some time in the 6th century B.C., when the Greeks became dissatisfied with mythic explanations. Gradually the Greek logos replaced the mythos of the earlier culture, and it is this logos that profoundly shaped the Greek mind, and indeed the nature of western culture since. The Graeco-Roman cycle comes to an end some time in the 5th century A.D. From the ruins of the disintegrated Roman civilization emerges a new form of logos, which I call the medieval *theos*, that in turn begins to shape culture, knowledge, and the nature of man's mind. We are entering the medieval cycle, which comes to an end in the 14th and 15th century, when the whole formation disintegrates, and the new form of mind embodied in the Renaissance arises.

However, as vigorous, virile and original as the Renaissance was, it did not quite succeed. *The Renaissance did not produce its corresponding form of logos*, which would give birth to a new civilizational cycle. Only in the 17th century did a new form of logos emerge, which gave rise to a distinctly new culture. This form of logos that I call *mechanos*, has shaped the destiny of western civilization for the last three centuries. But the modern mechanistic cycle seems to be coming to an end; *Mechanos*, as the contemporary version of logos, is unable to sustain us and simply is crumbling under the weight of its own contradictions. We now see a new form of logos emerging, which I call *evolutionary telos*. Although it currently manifests itself in a tenuous way, it is already organizing the variety of our experience and of our knowledge.

Each cycle has been organized, shaped, and controlled by its specific form of logos. This logos, which represents the essential simplification of the unmanageable reality of the previous cycle, is a stunning manifestation of the mind's limitations and of its powers: since it cannot dwell in complexities, too intricate, the mind must impose its simplified patterns of understanding. In brief, the history of western civilization is the history of growing complexity, as represented by the ontological process of becoming, which, however, is punctuated by the infusion of simplicity, whereby the mind, by imposing a new order on reality, organizes it in a new and simplified way. The epistemological order is superimposed on the ontological order. Indeed, the ontological order simply mirrors the epistemological one. Let me use a diagram to represent the four cycles.

Every cycle of unfolding complexity is governed by its specific logos until the complexities of a given civilization become unmanageable and the formation breaks down. A new logos, i.e., a new form of simplicity, emerges and organizes a new cycle, until it strangles itself on its own complexities. This is now happening in the cycle of *mechanos*, and its mechanistic civilization is breaking under the sheer mass of its complexity, resulting in incomprehension and confusion. We are at the dawn of another cycle, of another logos, of another form of simplicity, *evolutionary telos*.

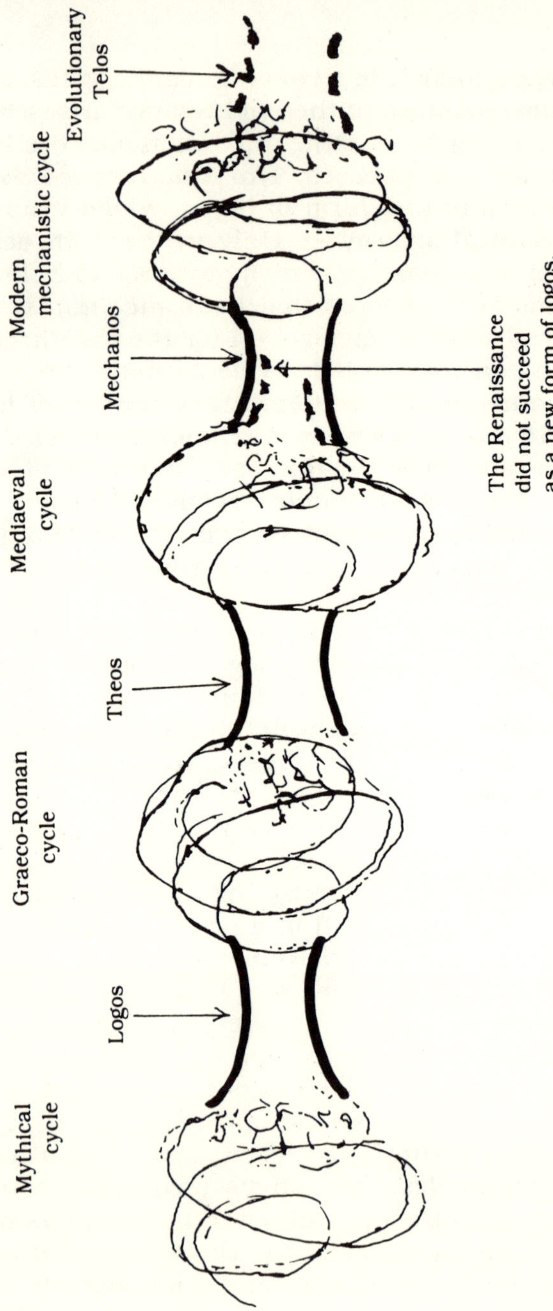

Figure 4–5: The history of Western civilization. Diagram represents the four cycles.

New insights and new extensions of knowledge are visibly piercing the walls of the mechanistic cosmos. Although it is hard for us to see, those insights are already forming themselves into a new pattern of simplicity, based on the cosmological unity of mind and reality, as well as a new form of integration between man and his universe.

In this section I have tried to provide an epistemological structure which attempts to accommodate the many insights of the New Physics, as well as many insights of other people seeking a non-physical paradigm in other branches of knowledge. We are approaching a new cycle, or to put it in more fashionable terms, a new paradigm of *understanding* which, when it is completed, will render to us a new cosmos. For a while, the outer walls of the cosmos and the inner walls of the mind will coincide. But only for awhile. For in times to come new insights again will pierce through the walls of the cosmos of our new evolutionary unity, and we will find that yet another form of logos will be needed to organize anew our knowledge and experience, or sense of freedom and the meaning of human life. Such is the story of this continuously unfolding universe of ours.

Mind—Sensitivities—Reality

Lurking behind the theory of the interactive mind is a new concept of man, who is defined as the subtotal of all his sensitivities. The mind is not a disembodied organ, working like a computer within amorphous flesh that otherwise has nothing to contribute to our thinking. Our flesh is thinking. So are the billions of cells within our body. Through this myriad we receive the singing of the cosmos. The more sensitive the ear, the more music it can receive. The more sensitive the organs, and the greater their variety, the more we can elicit from the universe.

Man has been defined as a rational being; as a political being; as *homo faber*. Yet what is most important about man is that he is a sensitive being, that he is endowed with sensitivities, that via sensitivities he is making his evolutionary ascent. Man is, in short, self-sensitizing, or self-transcending and thus a self-perfecting being. New sensitivities are new windows that enlarge the horizons of our world; they are also the vehicles by which we carry on the evolutionary journey and through which we make ourselves into more human and more spiritual beings.

When the first amoebas emerged from the primordial organic soup, they were victorious because they acquired a new sensitivity enabling them to react to the environment in a semiconscious manner, which was the beginning of all learning. For learning is a capacity, a sensitivity, to react in a feedback way to the environment and its conditions. The glory of evolution starts when organisms begin to use their capacities, their sensitivities, in a conscious and deliberate manner to further their well-being.

From the organic soup via the amoeba to the fish; from the fish via reptiles to primates; from the primates via the chimpanzee to man—this has been a continuous and enthralling story of the acquisition and refinement of ever-new sensitivities.

When matter started to sense and then evolved the eye (as the organ of its new sensitivity), this was an occasion of great importance, for reality could be now seen, could be articulated according to the power of the seeing eye. Without the eye to see, no reality could be seen. The eye brought a visual aspect to reality. One cannot exist without the other. For what is the seeing eye that has nothing to see? And what is the visual reality that has never been seen?

The seeing eye is a form of sensitivity through which we articulate reality around us. It is one of many, all products of the articulation of evolution, and more than passive repositories of the evolutionary process. Through them we process, apprehend, and articulate what we call reality. There is no more to our reality than our sensitivities can render. The emergence of every new form of sensitivity is a new window on the world.

With new sensitivities we articulate the world in new ways; we elicit from the world new aspects. The power of sensitivities is the power of co-creation. No aspect of reality imposes itself on us with an irresistible force. We absorb and assimilate it only when we acquire a way of seizing and comprehending it, when we come to possess an appropriate sensitivity.

This power of articulation is the power of creation. Nature goes about its business by endlessly articulating, and such is the story within the human universe. By acquiring new sensitivities we acquire new powers of articulation, and thus new powers of creation. Sensitivity, therefore, holds the key, not only to our evolutionary ascent, but also to the variety of our mind, and ultimately to our understanding of all knowledge.

Let us return to our earlier story of the evolutionary ascent. When the first amoebas started to articulate themselves from the original sea of the primordial organic soup, this was at once a triumph of life ascending and a triumph (still muted at the time) of consciousness arising. Amoebas started to *react* to the environment in a deliberate and semi-conscious manner. From this point on, the evolutionary tale is one of augmentation of consciousness and the continuous acquisition of new *sensitivities* through which organisms react to the environment in ever more knowing and purposeful ways. As their sensitivities multiply, organisms elicit more and more from the environment. They draw from reality in proportion to their ability to receive it and transform it. Here we can see that their reality was outlined by the nature and scope of their consciousness and their sensitivities; reality was outlined by the nature of their mind.

There is, therefore, an intimate relationship between our total evolutionary endowment in terms of consciousness, and all the knowing powers we possess, and the nature of the reality we construct, receive, and recognize. We simply cannot find, see, or envisage in reality more than our senses, our intellect, our sensitivities, our intuition (and whatever other evolutionary endowments we possess) allow us to find and see. The more sensitive and knowing we humans become, the richer and larger becomes our collected reality. What is beyond the mind of the species may be reality *in potentio*, but it is not reality as we know it. Our concept of reality *is* reality as we know it.

Upon receiving any aspect of reality, the mind always processes it, and, in processing, actively transforms it. Let us reflect on the meaning of the two expressions, "processing reality" and "transforming reality." Both are fundamentally inadequate. For they suggest that there is such a thing as an autonomous reality "out there," to which the mind applies itself and on which it works. Such a picture is fundamentally misconceived. There is no such thing as *reality as it is*, which the mind visits and on which it works. Reality is always intertwined with the mind that comprehends it, in the very act of comprehension, which is at the same time the act of transformation. We have no idea whatsoever what reality could be like *as it is*, because always,when we think of it, when we behold it (in whatever manner) *reality is invariably presented to us as it has been transformed by our cognitive faculties.*

Reality for the amoeba has been something less than that for the fish, and still less than that for the human being. The richness and diversity of the experience of reality is in proportion to the organism's capacity to receive and decipher it, to emphasize the point. The more primitive the capacities, or, to put it another way, the more primitive the mind, the more primitive the experience of reality. The more versatile and exquisite the mind, the more versatile the reality, and the richer the experience of it. Paradoxically, the organism receives from reality as much as it puts into it.

The interactive theory of mind is not an expression of old-fashioned idealism, which denies or mystifies reality. It is rather an expression of *suprarealism*, accounting for all the stages of the real in its evolutionary unfolding: it makes sense of the reality of the amoeba, of the reality of primitive cultures, of the reality of the scientific culture, of the reality of esoteric traditions.

The "objectivized" mind is one specific crystallization of our evolutionary journey. The scientific view of reality is one specific method of reality's reception and transformation. Yes, the reception of reality in the scientific key is its transformation—through numerous filters science places between us and the real out there. Some of these filters, such as Schrodinger's equations and other mathematical symbols, are sophisticated indeed.

The interactive mind is the expression of the total of the sensitivities that went into its making. These sensitivities are filters through which the mind views and receives reality. Mind is within reality and, at the same time, of the reality that surrounds it. Like the light within, it illuminates reality; by illuminating, it articulates; by articulating, it co-creates.

Let us represent:

Mind I—mind in the narrow sense; Mind II—the subtotal of all sensitivities of which Mind I is made; and Mind III—reality, within which both Mind I and Mind II reside—by three con-

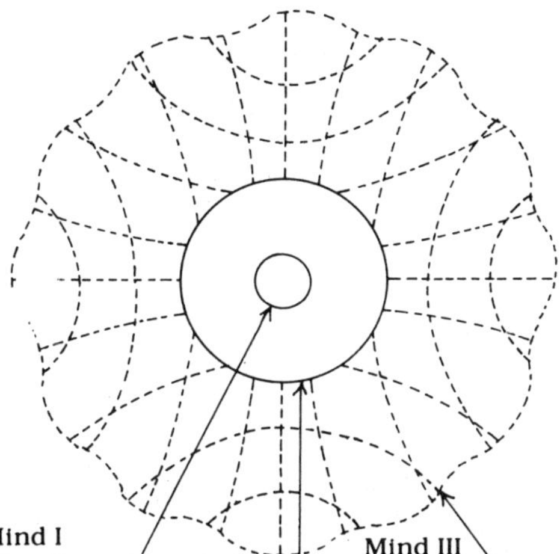

Figure 4–6: Three spheres merging into each other convey the essential evolutionary unity of our cosmology.

centric spheres (sphere I; sphere II; sphere III; of which each is merging with the next, and of which each is an aspect of the other two (see Figure 6).

To say that reality is a form of mind may strike some as an unjustified semantic liberty, but it is not. As mind is part of the real, so reality is part of the mind.

Sphere I and sphere III are aspects of each other. Consider the statement: "The symmetry of this pale-pink carnation is a beautiful phenomenon." Insofar as our carnation is a phenomenon, it belongs to the sphere of reality (sphere III). Insofar as it is "symmetrical," "pale-pink" and "beautiful," it belongs to the sphere of mind (spheres I, II). Insofar as we make a verbal utterance about it, it belongs to the sphere of the discursive mind (Mind I). Insofar as we appreciate its total beauty, it belongs to the sphere of sensitivities (Mind II).

Sphere III (reality) is "mind," but not in the same sense in which sphere I is. However, within the sphere of reality itself we can hardly find two different things which are real in the same sense.

The diagram, showing the three spheres merging into each other, means to convey the essential evolutionary unity of our cosmology. We are emphasizing here evolutionary and cosmological unity, but not homogeneity. Mind I and Mind III are aspects of each other, but at the same time exhibit different stages and different modes of the articulation of evolution.

In this essay I have addressed myself to the role of the mind in the universe of becoming. The becoming of the universe is inseparable from the becoming of the mind, an insight in perfect harmony with the new ontological vistas unveiled to us by present particle physics, which has abandoned the rigid, deterministic Newtonian framework and come to recognize that on the ultimate level of analysis the observer and the observed merged inseparably.

Let me put the matter in general terms: there is no objective reality in the absolute sense because there is no such thing as objectivity independent of our cognitive faculties. We do not photograph this (purportedly objective) reality in our scientific theories. The whole process, which we break into three component parts—the reality out there, the process of photographing it, and the process of developing the negatives, that is to say the process of the formulation of theories and of knowledge—are aspects of each other and cannot be torn from each other; they co-define and co-constitute each other. The nature of our mind is the nature of our knowledge, is the nature of our reality.

I am aware that I am proposing some rather uncomfortable notions. But the history of modern science has been the story of uncomfortable *and* improbable notions that have subsequently been built into the floor of our understanding of "reality." Ours is the universe of becoming: the universe of emergent qualities, the universe of new forms of understanding, which, although they at first appear as conceptual shocks, after a time resolve themselves as illuminating new insights into the nature of things.

The emergence of the mind is one of our lasting mysteries. Yet to conceive of the mind as a mystery is less mysterious and less mystifying than to fathom it as a mere brain working according to physical, mechanistic laws alone, for then that mystery is merely part of the natural order of things. Mind and imagination are bound together, and the nature of imagination gives us a far better clue to the understanding of the mind than a hundred neurophysiological studies. But the nature of imagination is wonderfully mysterious, as is the quality of the universe that mind/imagination renders to us. The unfolding universe is at the mercy of our mind/imagination for *it is* our mind/imagination.

> *But to the eye of the man of Imagination*
> *Nature is Imagination itself.*
> *AS MAN IS, SO HE SEES.*
>
> William Blake

All knowledge is mind. Epistemology, or the theory of knowledge, is the articulation of the modes through which mind manifests its knowledge. Those manifestations signify the peculiar human presence, which is cognitive, sensitive, sensitizing, processing, transforming. In short, the co-creative mind is the most important mode of our being in the process of becoming.

The story of the human mind is the story of liberation, and although we liberate ourselves from old constraints, we find that the process of liberation is never finished. So we have liberated ourselves from logical forms, from determinism, from the notion of absolute laws of science. We have been building a new paradigm of understanding positing such notions as the implicate order, dissipative structures, morphogenetics fields, participatory universe/mind, though we have not integrated these new posits into one homogenous, holistic pattern.

But one thing is clear. For perhaps the first time since Parmenides, and certainly since Kant, we see the mind as a marvellously creative instrument responsible for the shape of reality around us, molding the real in its own image, and promising us even more intoxicating freedom as we unlock new resources of our imagination. The interactive mind here proposed is different from Kant's conception of a rather rigidly structured mind. The whole agency of the mind must be seen as continually evolving, not as a fixed set of categories but as a myriad of possible patterns, guided by the depth of imagination, which ventures where even angels fear to tread.

The haze we see covering the universe is the haze of our minds. We have to make more of ourselves in order to make more of the world. We will have to see the mind much more clearly before we see the universe more clearly. We always redefine ourselves when we redefine the world around us, and vice versa. This continuous oscillating self-referential process is a part of our perplexity and our intellectual agony. We are the instruments on which the cosmos plays. We are, at the same time, the audience which listens to the symphony played by the cosmos on the instrument which we are ourselves.

Discussion

Henryk Skolimowski

René Berger: I liked your paper very much, but I would have liked to have had more precise information. It seems to me that you spoke about the mind as co-extensive with the universe. But I was expecting you to talk to us about the participation of the mind with the universe.

Henryk Skolimowski: The new physics gives numerous examples of the participation of the mind with the universe. You can refer to articles by John Wheeler or by Fritjof Capra. I preferred to elaborate on the philosophic aspect because I am a philosopher, and I tried to express, as you indicate, not only this type of co-extensiveness of the mind with reality, but the fact that reality, the idea that we have of reality, is inseparable from our mind said to come from it.

A Member of the Audience: I was sorry that in your paper you only dealt with purely theoretical notions, and in fact, epistemological notions; I would have liked you to tell us more precisely how all these neurophysiological notions that we have about the brain fit into your exposition. You want to show us that reality is mind. Could you not give us some more precise information, showing how your paper is connected to more precise ideas, such as neurophysiological ideas, for example?

Henryk Skolimowski: It is true that what I have proposed is only a model, and even a simplified model. But I think that if this model has elements which correspond to the brain and to that which the brain is, it should be able to be used afterwards by neurophysiologists in their study of the brain.

A Member of the Audience: I agree with the "interactive" mind in the universe, as you describe it, but I would like to understand a little better what supports your certainty that there is nothing to the mind except the knowledge that you have of things. Really, I would like to know if behind this knowledge that you have of things, which you assimilate with the universe itself, there is not something else on which your knowledge rests and whence it is derived. It seems to me that there is a reality behind your own mind.

Jean Charon: Yes, Henryk, you do speak in your paper about reality "in potentio." What do you mean exactly by that? Would it not be something fundamental on which your mind would operate: that is, that from this "potential" reality that we know nothing about, your mind would try to make an image of the universe that surrounds us.

Henryk Skolimowski: I am trying to ask myself questions about the nature of the imaginary. What is there in our imaginary that could be directly linked to reality, and what is there in this imaginary that comes directly from the imagination? In other words, is there an imaginary that can exist without any relation to reality? If we could differentiate between these two, then I think I could tell you what I mean by "potential" reality —reality *in potentio.*

5.
Externalization of Knowledge and the Humanization of Man

Jerzy A. Wojciechowski

Humanity and Evolution

Humanity is an unfinished business. It has been evolving since its beginning and it continues to evolve. Moreover, its evolution is constantly gaining momentum; it speeds up and becomes more profound. This means that at no moment in its history has mankind achieved its full development and realized its full potential. Advancement, development and progress are highly esteemed notions, but they create a serious problem as far as the knowledge of the human species is concerned. Namely, as long as mankind evolves, no description of it can be considered definitive. Consequently, we cannot expect to have a definitive knowledge of human phenomenon. The more we evolve, the more we will be surprised at ourselves.

Humanity has not actualized its full potential; it has not come close to doing so. Quite the contrary. The more it develops, the more it increases its capacity for action and for further development. Its abilities grow in size and in power. There are constantly more people, more knowledge, more rational activity, and greater possibilities for advancement for individuals, societies and mankind as a whole. These are the properties of an eminently open system. For the sake of clarity, let us bear in mind that open systems are those whose very nature implies exchange, a give-and-take relationship with the environment. Consequently, an open system is necessarily a dynamic, evolving entity.

All living beings are open systems, not only individual organisms, but the whole hierarchy of living species. This does not mean that openness is always the same throughout the living kingdom. The degree of openness varies from individual to individual within the same species, and the differences increase when individuals of different species are compared. The interspecific differences are not accidental. They increase with the evolutionary progression of forms of life: the more advanced the form, the more open the system, i.e., the greater is its potential for development and intraspecific variety. This fact may be expressed in the form of a law:

> *Law I: The openness of a living system is proportional to its biological perfection.*

This means that a living system's openness is proportional to the position it occupies in the evolutionary hierarchy.

All life evolves, but as we have just intimated, all living beings do not evolve in the same way nor to the same degree. Human evolution in particular is unlike that of other species. What makes it so special is the activity of the intellect. Because of it, man is involved not in one kind of evolution like other organisms and species, but in two kinds, namely material and intellectual, or biological and cultural. These two types of evolution are as different from each other as the intellect is different from the body and the intellectual activity is different from the physical activity of the organism. Neither one can adequately account for or explain the other.

Of the two, biological evolution is more basic, much older, and more universal. As far as we know, it results from three processes, namely the apparently random generation of new forms, the elimination of the unfit variations, and heredity. Nature is quite prodigious in the production of new variations, but quite wasteful as well. Only a small percentage of new forms survive the test of adaptation. In terms of human time and compared with the rate of cultural evolution, biological evolution is slow, but nature is in no hurry. If cultural evolution were to progress at the rate of biological evolution, humanity would not yet be past the stage of *australopithecus* in its development.

To understand cultural evolution, it is necessary to have an idea about evolution in general. Since the consequence of evolution is the generation of ever more perfect forms of living beings, it is plausible to conclude that the business of evolution is the advancement of life. Whether the expansion of life is the product of blind forces and chance events, as some schools of thought believe, or whether it is a purposeful process, the fact remains that as a result of evolution there is more life in general. Not only is there a quantitative growth of biomass but, moreover, life becomes more varied, more complex, more powerful and more efficient. In order that there be a progression of life forms, the new forms that survive elimination and become established elements of the tree of life must in some way be better than the older ones. To be more advanced, they must have a greater potential for existence. They have to be able to be more active, to do more varied, more complex and more difficult things. Moreover, they have to have a greater life capacity, which in the first place means a greater probability for their offspring to develop and survive to the reproduction age and to reproduce, i.e., to continue and to expand life (Table 5–1).

Evolution may be described as the generation of devices producing increasingly more advanced forms of energy and using them more efficiently. The hierarchy of non-sentient (vegetal), sentient, and rational organisms reflects this progression. For organisms, the payoff for better handling of energy is greater life-worthiness. Seen from the point of view of the capacity for life, rational creatures are much more advanced organisms than irrational ones. Humans have the highest probability of survival and reproduction among all living species (Table I). They can make the most efficient and economical use of life energy. Their efforts can survive them and contribute to the life of subsequent generations. All organisms are biodegradable and after their death may serve as fertilizer for their descendents or other forms of life, but they do not contribute otherwise in any perceptible way to the well-being

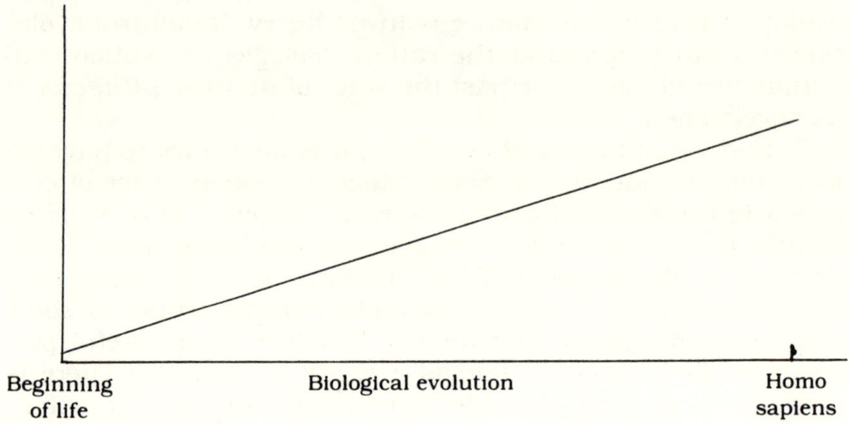

Table 5–1: Chances of survival of the offspring. Beginning of life, biological evolution, homo sapiens.

or evolution of future generations. Human bodies degrade like any other organisms, but their ideas are not biodegradable. Written down, they survive and contribute to the building up of the sphere of culture, which influences subsequent generations in a decisive way.

Not only are humans the most advanced handlers of energy, but with time they become better and better. The more they progress, the more they learn how to protect themselves and to increase their chances of survival (Tables 5–2 and 5–3). There are therefore reasons besides rationality that justify assigning man the highest place in the evolutionary hierarchy. The other reasons are, however, intrinsically related to rationality.

Because humans are so different from irrational creatures, one may wonder whether they do not contradict the rules of the evolutionary process. The appearance of the thinking creature was a radical novelty, and the birth of self-reflexive consciousness was an event of cosmic importance. However, the becoming of the human species nevertheless did not represent a break with the principle of evolution; to the contrary, it was a supreme fulfillment of that principle. The new species was an advancement, and a considerable one, over existing spe-

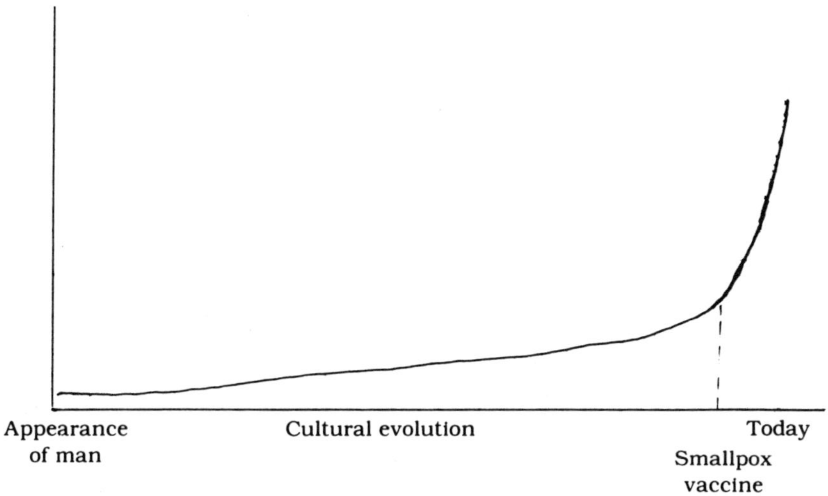

Table 5–2: Chances of survival of the offspring. Appearance of man, cultural evolution, smallpox vaccine.

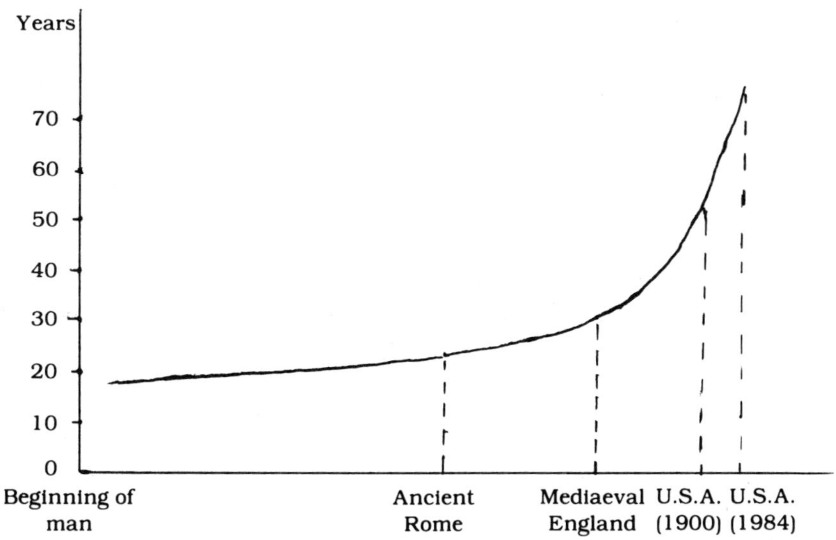

Table 5–3: Life Expectancy. 1900-in U.S.A., 1984-in U.S.A.

cies. It did not negate the preceding forms of life; it incorporated their mode of existence in a more perfect form and gave them a more eminent expression. With the appearance of the human being, the advancement and experience of life developed during the course of biological evolution was not lost; it did not come to naught. Instead, with man, the evolutionary process became faster and more complex.

Externalization of Knowledge and Evolution

As mentioned earlier, biological evolution cannot account fully for cultural evolution. The Cro-Magnon man and the Nobel Prize winners belong to the same species; their differences are due not to biology but to cultural evolution. Different consequences indicate different causes, and the "mechanism" of cultural evolution obviously must be different from the "mechanism" of biological evolution. All evolution is coevolution; nothing evolves alone. It takes two to evolve: an organism and its environment, related to each other by a feed (feedback, feedforward) relationship. Without this relationship, which involves mutual influences, there would be no incentive for organisms to evolve. Evolution, whether biological or cultural, is a relational process. In this respect these two types of evolution are similar.

There would be no evolution without accumulation, preservation, and transfer to consequent generations of successful morphological variations and experience. This is true for both types of evolution. The difference between them consists, first of all, in the kind of experience in question and, secondly, in the mode of their accumulation, preservation, and transfer. In the infra-human realm, experience is the product of exchanges between organisms and the environment composed of other organisms and inanimate matter. On the human level, experience is twofold—organic and intellectual—and so are the exchanges which underlie it. Not only is human experience incomparably richer, but it is stored and preserved differently than in other organisms.

Irrational creatures store and preserve information and experience within their organisms. They cannot do otherwise; there is no body of animal experience existing and persisting outside of their organisms. Consequently, the capacity of the single organism to store experience is the total amount of experience stored. This is why the amount of experience within a given species is stable and does not grow with consequent generations. It is proportional to the storing capacity of the organism, i.e., it is species specific. The only way for nature to increase the storing capacity and the level of experience is to produce a new kind of organism. The question is what reason, if any, is there for doing this. The only reason that we can think of seems to be the pressure of the environment and the need of better adaptation to and coping with that challenge. If this, indeed, is the case, then it is plausible to infer that the generation of random variations is the organism's way of searching, under the environmental challenge, for a greater potential for experience and thereby for "life worthiness."

As we mentioned earlier, humans experience and store experience differently from other animals; this difference accounts for the disparity in the modes of life and evolution of humans and irrational creatures. That human experience is richer and more varied than that of other animals is well known. What is less well understood is the difference in the storing of experience and the consequences of the specifically human mode of storing it. Humans store experience in not one, but two different fashions, inside themselves (organically) and *outside* their bodies. The latter mode of storage allows cultural evolution and elevates man to his present level of development.

In order to understand how the twofold mode of storing experience could develop, it is necessary to analyze and compare the two levels of human knowledge, namely sense perception and intellectual knowledge. From the point of view of the storage of knowledge, the principal difference between sensation and thinking consists in the manner they can be made known to others, i.e., communicated. Sense knowledge can be manifested directly through body language or sounds, as animals do. The dance of a bee indicating the direction and the distance of flowers, the barking or waggling of the tail by a dog are examples of the manifestation of sense knowledge. This mode of communication is limited a) in range—by the sensing capacity of the animal, and b) in time—only simultaneous, not later, communication is possible. The message cannot be stored outside of the organism.

Thinking can be expressed in two ways: it can be manifested in a limited way through body language or it can be externalized by speech. Speech makes all the difference, as far as communication is concerned. First of all, speech allows a much greater richness of elements of communication than body language: the number of words can grow indefinitely, in contrast with body signs. Second, and more importantly, words can be preserved in time. They can be written down and stored outside of knowers. The ability to store thoughts expressed in words outside the knowers opened the possibility of transcending the limits of memory and of indefinitely increasing knowledge. This growth of knowledge, though of fundamental importance of mankind, is only one of the consequences of inventing external storage of knowledge.

Intellectual knowledge produces two kinds of effects: intellectual and material. Each act of thought, abstract and theoretical as it may be, produces consequences beyond this act itself. In the first place, it changes the state of the knower, who becomes more knowledgeable as thinking exercises his intellect and makes it more capable of intellectual effort. This important fact may be expressed in the form of a law:

Law II: Thinking increases the potential for thinking.

This, however, is not the only consequence of thinking. Thought is power, therefore:

Law III: Thinking increases the potential for doing.

Communication expands the power of thought. It disseminates knowledge, produces involvement and cooperation. The thinking of the individual knower enriches society, just as the body of knowledge possessed by the society enriches individual knowers. It is therefore plausible to state that each externalized act of thought changes the state of mankind. Whether they realize it or not, humans are engaged not only in the process of personal devleopment, but also in the evolution of their species.

The relationship of the rational animal to his species is different from that which exists between individual nonrational animals and their species. Sheldrake's hypothesis notwithstanding, nonrational animals do not conceive the idea of progress and do not act consciously to improve the condition of their societies. Their brains are not a change-engendering organ; the human brain is. This is why man's impact on his species and its development is incomparably greater than that of other animals on their species. Moreover, the more humans know, the more they infleunce their life conditions and increase the rate of the evolution of mankind. This important fact can also be expressed as law:

Law IV: Man's capacity to influence the development of mankind is proportional to knowledge.

The accompanying chart (Table 5–4) represents the consequences of the externalization of knowledge. The key distinction is between direct and indirect communication. It expresses the difference between the storage of knowledge exclusively in knowers, i.e., in the memory of individuals, and the storage outside of knowers. The rest of the chart describes some of the consequences of the two modes of storing knowledge. Of course, the chart should not be considered an exhaustive description of the situation, the complexity of which is still beyond our full grasp, but some general conclusions can be formulated concerning the development of man.

The more humanity progresses, the more there are knowers, the greater becomes the cumulative brain capacity and the greater the body of accumulated knowledge. Thus the advancement of mankind generates increasing capacity for further evolution. The more humanity evolves, the more it can and does progress, the more it becomes the cause of its own development. Biological evolution is a product of nature, i.e., of the supposedly random interplay of nonrational factors. Cultural evolution, i.e., the humanization of man, is a progressively rational and conscious process. Through his intellect and knowledge, man becomes increasingly his own maker and product. The humanization of man is the crowning achievement of evolution, but it is not an unmitigated success.

Table 5–4: Externalization of knowledge and its consequences.

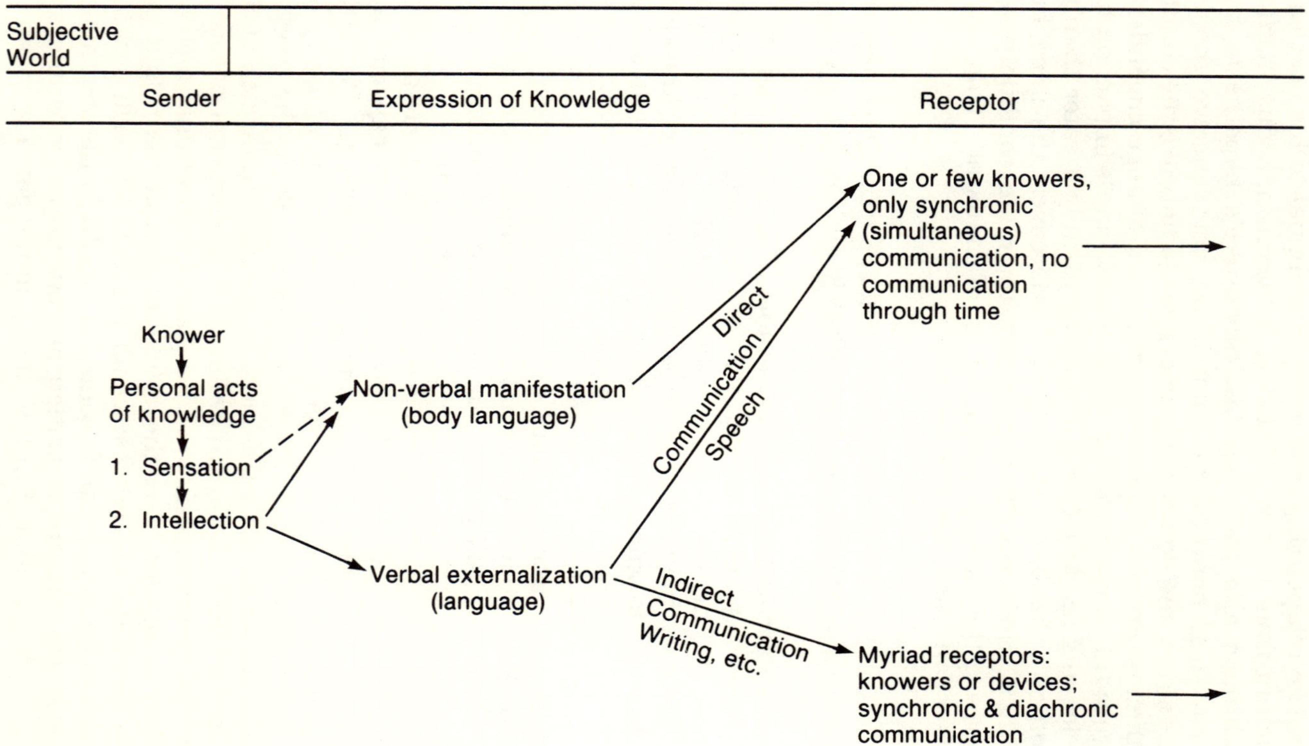

Objective World					
Storage of knowledge	Consequences				
	Knowledge	Culture	Population	Social Organization	Relation to Nature
Storage in knowers Consequences: 1. No storage outside of knowers 2. Storage limited by the capacity of memorization 3. Limited retrieval 4. Memorization, retrieval—subjective acts 5. No verification of these acts or contents possible 6. No science	**Knowledge Construct I** Consequences: 1. Not cumulative beyond the limits of memory 2. Stable: capacity for growth limited by the limits of memory	"Stable state," non-literature cultures: hunter, gatherer, nomadic, primitive agriculture	More or less: sparse, steady populations	1. Small (organic) groups: family, band, tribe 2. Principles of unity & organization: a) procreation, b) lineage, c) authority of the principal procreator. General characteristics: stability of social structures 3. Relationships: a) diachronic: common ancestor, b) synchronic: personal, one to one 4. Status of the individual: no insistence on individuality apart from group identity. Basic unit: the group 5. Property & law: only private (ruler's), no public property or law 6. Work: little or no specialization among adult males or females	Relatively non-destructive, in ecological balance with nature
Storage outside of knowers Consequences: 1. Unlimited storage capacity 2. Unlimited retrieval 3. Storage, retrieval—objective acts 4. Verification possible 5. Science	**Knowledge Construct II** Consequences: 1. Cumulative 2. Capacity for growth unlimited	Progress Cultural evolution leading to world culture	Accelerating growth	1. Large and growing societies 2. Principle of unity: culture territory, common past, Principle of organization: varied, no universal model. General Characteristic: variability 3. Relationships: a) diachronic: common social (national) history & culture, b) synchronic personal & impersonal 4. Status of the individual: greater distinction between individual & group. Basic unit: group or individual 5. Property & law: private & public 6. Work: specialization	Destructive, excluding man from the system of nature's ecological balance

Discussion

Jerzy A. Wojciechowski

Willis Harman: If the development of gnosticism that Gilbert Durand told us of is about to arrive, and if it becomes as important as he thinks it will be, it is going to alter incredibly the picture you have just given us. To Durand a completely different knowledge is appearing, which will be centered around gnosticism, around unification apparently very far from an incredible development of electronics. And this will greatly affect human behavior.

Jerzy A. Wojciechowski: Yes, you are absolutely right. What is clear to me is that we cannot continue the development of humanity as we have been doing over these past few decades. We must develop a new form of wisdom. The Greeks distinguish between knowledge and the "Noûs," between rational knowledge and intuition, and at present we absolutely need this intuitive knowledge. We need to integrate all the knowledge we have at this moment, but this will not eliminate the problem of the relationship between humanity and its knowledge. This will not stop us from being aware that we are faced with a growing body of knowledge that will continue to grow unceasingly, and that from now on no one can master such knowledge completely.

Willis Harman: If you were as anxious as I am to see this new form of intuitive knowledge appear, then I think you would have modified your paper with respect to the first version you gave at our preliminary meeting last spring, when I made the same remarks.

Jerzy A. Wojciechowski: I was struck by your arguments, but we need this type of scientific knowledge, and not just intuitive knowledge, because there is a direct relationship between the level of humanity and the need for useful knowledge. If the level of humanity compared to its present level increases, then we need a scientific knowledge, if only to encourage humanity. The difference between humanity and one individual is that the individual can always opt for the desert, go there, and spend his time in contemplation. Humanity cannot do that. Society as a whole cannot become a society of mystics! So the biological needs of humanity must be met. This does not mean that each of us needs filet mignon, or a new pair of shoes every year, and so on. But we do need shoes, we do need to eat something, and these things must be produced. Contemplative knowledge cannot satisfy these needs.

Willis Harman: However, this could change our values considerably.

Jerzy A. Wojciechowski: Yes. Of course we can hope that this will change our values, but that could be the subject of another meeting.

A member of the audience: I, too, am a professor of philosophy, and I was extremely interested in your paper. However, it leaves several questions in the shadows, and I would like in particular to ask you the following: why do you think the West has developed such an unbalanced body of knowledge? I would like to ask you a second question too: knowledge is first of all the development of rationality and reason, and usually people are proud of their reason; so how can people, while using this reason, be led almost to destroy themselves? I will again mention the atomic bomb. It would seem that reason should protect humankind and protect rationality. Do you have any answer to these two questions?

Jerzy A. Wojciechowski: I will answer your first question. Why did such an unbalanced knowledge develop? I will limit myself to the West. I think that the West has for a long time been founded on a certain kind of rationality which we have chosen. And I see two sources of Western civilization: first, the rationality and logic of the Greeks, and next the Hebraic tradition, or more generally speaking, Christianity. I will begin with the Hebrew tradition. That tradition has developed an idea of man and especially his relation to nature. The essential notion in the relationship of man to the world in the Hebraic tradition is that this relationship is based on what is called objectivity. What does this objectivity imply? A separation between man and nature. Objectivity presupposes that we can look at the world from the outside, and from above. If I measure a table, I can do it objectively, because I consider myself distinct from the table. The Greek tradition also presupposes this objectivity, by the definition I have just given, but on this point the Hebraic tradition takes on special significance. We must return to Mount Sinai, when God was speaking to Moses. And because God spoke to Moses, that means that Moses and the Jewish people and man in general are clearly superior to all other creatures, because man is God's conversational partner. Thus man can look down from above on all the things in nature which are distinct from him, and this perspective is absolutely essential to the Western mentality. The Greeks, too, supposed that man, and especially the Greek man, was different from the others, and he can especially look down on the barbarians. The Greek considered himself unique in all humanity; he considered himself distinct from the barbarians, which means all people except the Greeks. Our Western science derives from these two sources. Our science must make of man a being able to think of himself as distant from all that surrounds him, able to look objectively, that is, "from above," on the world around him. That is the basis of Christianity, and Christianity founded western science. Science is thought to have begun as a reaction to religion: I think this is a completely inaccurate idea. Christianity is the matter of modern science, at least of Western science. And rationality, reason, is based on this notion of objectivity we have just discussed. This objectivity removes any notion of value from rationality. With science, I can look at things independently of any interest I could have in certain values. Science and values are mutually independent, and that is why science is going to be developed on the basis of quantity, while totally forgetting quality, which is linked to values.

In the West, we have developed another notion which is parallel to objectivity and which is almost a consequence of objectivity, and that is the notion of personality. Because man is superior to all the other things of creation, the person that is man is of great interest. Therefore we see the West looking at things from above, and pronouncing judgments, which are objective judgments, which thus do not take values into consideration. The West is developing perhaps excessively the notion of the person, of the personality of man, of the personality in each one of us.

I hope that this answers your first question about the reasons for the imbalance in western knowledge. In the face of the immense stature of technology, which is continually growing, the immense development of the quantitative, there is no corresponding development of the qualitative, that is, of values. And I think that this can explain how our Western civilization is now groping in the dark trying to find a ray of light coming from another civilization. In the great cities such as Paris, New York, Montreal, San Francisco, Los Angeles, you can see more and more people seeking gurus, spirituality, orientalism sometimes, and all this corresponds to an often unconscious need to open a window on the values of *other* civilizations than our own Western civilization. All these people are moving outside because they are looking for something that they can not find at home and that they feel, in a more or less confused way, that they need.

Now for your second question: Why does our beautiful rationality tend to lead us, in our Western world, to self-destruction? And you mentioned the atomic bomb. I personally witnessed the most terrible destruction of the second World War, that of Warsaw, in Poland, in 1943, and then again in 1944. There was no atomic bomb there, but out of one million three hundred thousand people living in Warsaw, 83 percent died; that is many more than in Hiroshima and Nagasaki combined. Thus, I know, I know in my very being, the power of destruction that man is capable of. Having lived through this, however, spurred me to reflect on man's power of destruction. I think that the challenge we have created, namely the challenge of *knowing* we can destroy ourselves, manifested by the existence of the atomic bomb, is really a necessary challenge, and probably a healthy one, perhaps the only one capable of saving humanity. We are usually terrified of the atomic bomb; we consider it a calamity, and it is, naturally, a frightening weapon forged by humanity and by human knowledge. Yet I think that the advent of the atomic bomb has a metaphysical meaning and even an evolutionary one. The atomic bomb forces us to become wiser. This is without doubt one answer to your second question, and that is why I think that humanity will be assured a future. Of course I may be wrong, but that is my answer in any case.

6. Consciousness as Causal Reality: Toward a Complementary Science

Willis W. Harman

> *. . . our normal waking consciousness, rational consciousness as we call it, is but one special type of consciousness, whilst all about it, parted from it by the flimsiest of screens, there lie potential forms of consciousness entirely different. . . . No account of the universe in its totality can be final which leaves these other forms of consciousness quite disregarded. How to regard them is the question—for they are so discontinuous with ordinary consciousness.*
>
> William James, 1902

> *Current concepts of the mind-brain relation involve a direct break with the long-established materialist and behaviorist doctrine that has dominated neuroscience for many decades. Instead of renouncing or ignoring consciousness, the new interpretation gives full recognition to the primacy of inner conscious awareness as a causal reality.*
>
> Roger Sperry, 1982

Ken Wilber, in his recent book *Eye to Eye*, documents impressively that mystics and philosophers in various traditions have consistently recognized three modes of acquiring knowledge: an "eye" of flesh, an "eye" of reason, and an "eye" of contemplation, as they are referred to by the Christian mystics. The first gives access to the outer, sensory realm; the second to the mental, symbolic realm; and the third to the transcendent, spiritual realm. Science is an activity of the mental realm; conventional science is the "eye" of reason looking to, and ordering the experience of, the outer realm. But there is a second kind of science, at least potentially—the "eye" of reason looking inward, and ordering the realm of subjective experience. This complementary science has been largely neglected by the Western world.

However, as Roger Sperry points out in his landmark paper on "Changing Priorities":

> Beliefs concerning the ultimate purpose and meaning of life and the accompanying worldview perspectives that mold beliefs of right and wrong are critically dependent, directly or by implication, on concepts regarding the conscious self and the mind-brain relation and the kinds of life goals and cosmic views which these allow. Directly or indirectly social values depend . . . on whether consciousness is believed to be mortal, immortal, reincarnate, or cosmos . . . localized and brain-bound or essentially universal. . . . Recent conceptual developments in the mind-brain sciences rejecting reductionism and materialistic determinism on the one side, and dualisms on the other, clear the way for a rational approach to the theory and prescription of values and to a natural fusion of science and religion.

Thus the Western neglect of the "complementary science" of inner experience has had serious consequences in our confusion about values. For it is ultimately in this realm of the subjective, the transcendent, and the spiritual, that all societies have found the basis for their deepest value commitments and sense of meaning. But how to deal with the subjective in a way true to the spirit of scientific inquiry, and yet not in the process to distort it that in the end the resulting science is no longer true to its subject? Aye, that is the question.

Throughout history, individuals and communities seem repeatedly to have come upon the creative factors and forces of the human psyche. Great philosophies and great religions have time and again come into being as an outcome of such discoveries, and for a time have profoundly influenced the course of human events. But as often as the discoveries have been made, they appear to have been lost. With the vogue of positivistic science in the earlier part of this century, the religious meanings associated with such experiences were rather thoroughly debunked, and serious exploration of the creative unconscious discouraged.

Recently there has been a resurgence of interest, however, both in the broader society in various meditative disciplines, mind-body approaches to health care, and religious philosophies, and also in the scientific community in research on consciousness. Scientists of considerable stature such as neuroscientist Roger Sperry, physicist David Bohm, and biologist Rupert Sheldrake have argued what would have been rank heresy only a decade or so ago: that there are unmeasurable, nonphysical aspects to reality that have to be taken into account because they are part of human experience—not the experience of the physical senses, but the experience of the deep intuition and of alternative states of consciousness. As William James said, "No account of the universe in its totality can be final which leaves these data disregarded."

The question which troubled scientists a half century ago, "Does mind exist?" (since it cannot be measured) seems strangely anachronistic today. The question remains, however, how to study mind in a way compatible with the scientific tradition of open inquiry and public validation of knowledge, and yet not be bound by too narrow and inappropriate a concept of "scientific method."

It seems apparent that further progress with this "second kind of science" will require a re-examination of the basic concepts of "objectivity" and "replicability."

Objectivity and Replicability

Consider the commonplace testing of the reliability of a new analgesic by giving the drug to a large population of subjects and making a statistical analysis of the data. In judging the merits of such research there would tend to be a shared understanding that pain is a subjective response affected by all sorts of subtle variables, such as the subject's desire to cooperate with the experimenter, the placebo effect, deliberate or inadvertent suppression of pain by autosuggestion, individual uniqueness, variability of the subject's responses from day to day, etc. Even with all the sophistication of double-blind techniques one would not expect, in research that so centrally involves self-reports of subjectively felt pain to be able to apply rigid definitions of "objectivity" and "replicability." Yet an outsider examining this research would not be inclined to be hypercritical of the results just because of this impossibility of demonstrating strict objectivity and replicability.

In research more directly related to exploring the powers of mind, problems involving objectivity and replicability become more acute. Take for example the phenomenon of psychokinesis, or of healing cancer through imagery and suggestion. In both cases we have careful work carried out by competent researchers; in both cases the results strongly suggest the existence of a genuine phenomenon. Yet because of the absence of some sort of "mechanism" to account for the phenomenon, stringent demands are placed on the research findings, and they fail to satisfy these strict standards. Researchers with positive beliefs about the phenomena tend to get positive results, and skeptics tend to get negative results—thus raising the question of objectivity. And the results are erratic even under what appear to be similar conditions.

The skeptical tend to interpret these characteristics of the research as evidence for the nonexistence of the phenomena, but they can also be interpreted as indications of inappropriate demands for "objectivity" and "replicability." Once one postulates the possibility of the subject's mind affecting the environment at a distance it becomes apparent that the state of the experimenter's mind can also affect the results—so strict objectivity is an impossibility. It is well known that human beings make choices at unconscious as well as conscious levels (*e.g.*, addiction, an unconscious choice overruling a conscious one; or guilt feelings, a message from part of the unconscious mind that it disapproves of a choice made consciously). But in psychokinesis or cancer self-healing, unconscious as well as conscious choices are involved; the person may want one thing at one level of the mind and something else at another. With this uncontrollable factor entering in, replicability is impossible.

But if these strict interpretations of objectivity and reliability through replicability are inappropriate in research on subjective experience, surely there are criteria something like these which *are* appropriate. Perhaps as the scientific exploration of these areas becomes more mature, something like the Buddhist ideal of "nonattachment" will replace the demand for strict objectivity. And something like "trustworthiness" (perhaps established on the basis of multiple imperfect tests) may replace strict reliability through replicability.

Consider as an example some of the recent work on multiple personalities. Is it really true, incredible as it may seem at first, that the same body may be inhabited by more than one personality? The ideal of nonattachment to preconceived notions of what can and cannot be is clearly pertinent. Trustworthiness of the psychiatrist's impressions of wholly different personalities expressing themselves through a single body is established by correlating these impressions with various body characteristics—voice patterns, physiognomic features, brain waves, and other distinguishing features. It turns out that when the personality shifts, as subjectively experienced by the psychiatrist or researcher, these various measurable body characteristics shift too, increasing the plausibility of the multiple personality model.

Characteristics of a Complementary Science

This latter example raises yet a different question. Most of conventional science seeks reductionistic explanations for phenomena. But "personality," like "health" and "love," is a holistic concept. How is truth sought in an area where the internal workings of the human mind are inevitably a part of the holistic situation? Probably it is essentially by carrying out explorations of the most profound and moving inner experience in an atmosphere of mutual trust and rapport, comparing experiences openly. (Contrast this with the deliberate deception of subjects that is typical of laboratory research in psychology). The researchers need to be dedicated to the search for knowledge but nonattached to specific assumptions, theories, and expected outcomes. The researcher must allow himself to be changed by the process of inquiry; he will be learning about his own consciousness in the process of studying the subject of consciousness.

There is some precedent for these conditions in the training of cultural anthropologists and psychotherapists. These disciplines involve learning to be free from judging a society or another person in terms of internalized cultural preferences, accepted mores, likes and dislikes, prejudices, and so on. In the process of coming to know another culture or another person in this intimate way, the investigator/therapist finds himself changing, sometimes profoundly.

There are other ways as well in which this "second kind of science" is intrinsically different from conventional science. Whereas the "science of the outer" is strongly positivistic, focusing on what is physically measurable, the "second kind" is potentially transcendentalist in outlook. The former tends to be deterministic; the latter, teleological, admitting of the concept of purpose. Conventional science has little to say about value issues; the "science of the inner" is inherently value-attentive.

We might summarize the contrast as follows:

Characteristics of Ordinary science	**Characteristics of Complementary Science**
Objectivity of investigator valued	Nonattachment of investigator valued
Replicability of results valued	Trustworthiness of results valued
Positivistic, dealing mainly with that which is physically measurable	Transcendental, inclusive of the nonmaterial and nonmeasurable
Reductionistic	Holistic
Deterministic	Teleological
Value-inattentive	Value-attentive
Production-and-control focused; emphasis on knowledge to generate technology	More emphasis on knowledge useful to guide human growth and development or to illuminate meaning

As an example to illustrate the contrast and complementarity of the two modes of inquiry, consider the phenomenon called "channeling." The term usually connotes a kind of communication to the conscious mind that appears to emanate from a source external to the individual's "ordinary" conscious-unconscious mind system, and it may seem in some respects to be quite superior to the ordinary mind. This "other" source may seem or claim to have been a person once, but no longer alive on this earth, or a being of some extraterrestrial sort not embodied in the usual physical sense. Other terms have been used for some forms of this phenomenon at other times or under specific circumstances—terms such as "inspiration," "creativity," "deep intuition," "mediumship," "religious experience," "revelation." The history of science, mathematics, art, and intellectual achievement is replete with anecdotal episodes of such communication from the "muse" or "genius" or "demon." (Our ambivalence toward all this is revealed in the two meanings given to the word "demon" in the English language.) The revealed scripture in many religious traditions are purported to have derived from such a source. Clearly the channeling phenomenon, however interpreted, has profoundly affected the history of the human race. The deeper value commitments and meanings found in most societies trace their origins to some such communications.

Yet ordinary science has had very little to say about the phenomenon, beyond admitting that most of its conceptual advances seem to have come from such a source. When something like this has been studied at all, it has typically been in the context of pathology. Let us explore how the subject might be dealt with in the context of "a second kind of science."

The metaphor of alternate personalities appears to be fruitful. Multiple personalities have been studied fairly thoroughly as a pathological occurrence. In such cases the usual personality appears to disappear and be replaced by another, often quite different one. The transition is typically, though not always, not under the deliberate conscious control of the person. More than one alternate personality may be involved. This personality may be quite dissimilar from the usual one in speech and thought patterns, mood and temperament, and other personality characteristics. But more than this, as noted earlier, there may be significant and even startling differences in voice characteristics, apparent gender, accessible memories, physiognomy, posture and movement patterns, brain wave patterns, chemical composition of body fluids, allergies, and other physiological characteristics—leaving little doubt that in some important sense the alternative personality "really exists" as surely as the usual personality. To be sure, it would be difficult to define or measure either one, but nevertheless a number of meaningful research questions can be asked, for example, regarding persistence of the alternates, conditions under which two or more personalities may "fuse" into one or alternate personalities may "leave" correlations between psychological and physiological factors, and so on.

Once the existence of multiple personalities in the pathology context has been widely recognized, it is a minor leap to recognize that some of these alternate personalities may be—as the subject often believes—knowledgeable and wise beyond the capabilities of the person's ordinary mind, and beneficent as well. Thus the characteristics of the alternate personality can be explored, and various research questions investigated, *while leaving open metaphysical questions* such as whether the source is "really" the same person who once was embodied in a life on earth, or is "really" some sort of extraterrestrial, other-dimensional, or spiritual entity.

Research questions of interest might include:

> What are the characteristics of the source of the alternate personality? How does the alternate compare with the main personality in intelligence, knowledge, wisdom, etc.?
>
> Does the alternate typically "take over" the physical body, or remain in the background and communicate with the main personality, who continues to hold center stage? If the alternate does "take over," are physiological characteristics altered?
>
> What factors influence the appearance of the alternate? To what extent is this appearance under the control of the main personality?
>
> Does the alternate appear to have access to knowledge that would not be presumed available to the main personality?

How could one go about verifying whether or not the source or alternate personality is who he/she claims to be?

It is important to carry out such explorations and to learn from them what can be learned. Such research will surely take us far beyond our present state of understanding, and will reduce the tendency to dismiss the channeling phenomenon as of no importance because scientists do not study it. Nevertheless, it is important to recognize that we eventually encounter some questions that are indeed beyond science, even in its extended form—although not necessarily beyond ultimate human understanding.

Suppose, for instance, that we were to conclude that in some cases of channeling there is contacted a source which appears to far exceed ordinary humans in intelligence, knowledge, and wisdom. Our tendency to ask of everything what it can be "used for" may have to be curbed in this case. For attempting to exploit such a source would be like the rat in the laboratory seeking to perform an experiment on the scientist. A far more humble attitude is appropriate if we are to learn from the universe what it would teach us.

The Hypnosis of Cultures

One of the most powerful implications to come out of the study of subjective experience derives from well-known phenomena in the field of hypnosis. The way we perceive the world and experience reality can be dramatically altered by hypnotic suggestion. Perception is shaped by the unconscious belief system, and one of the ways it can be temporarily altered is through accepting the suggestion of the hypnotist. Through hypnotic suggestion a person can be led to perceive something that is not there, or fail to perceive something that is there. The suggestion of a solid wall becomes so real to the subject that his fist is bruised upon "striking" it. The suggestion that a pencil is a hot soldering iron causes it, when laid on the back of the hand, to raise the physical signs of a burn. Acceptance of the hypnotic suggestion that one cannot lift a light object produces a complete inability to do so. On the other hand, a more positive suggestion may lead the body to perform feats it could not otherwise do—form a rigid bridge between two chairs, for example, or lift a heavy weight.

Each of us is subjected, from infancy onward, to a complex set of suggestions from our social environment, and they in effect teach us how we should perceive the world. We may from time to time, especially in early childhood, have experiences that do not conform to this cultural perceptual norm, but we eventually "correct" these perceptions and through the power of the socializing process cease experiencing these anomolies. Each of us is literally hypnotized from infancy to perceive the world the way people in our culture perceive it. Persons brought up in a very different culture might perceive it very differently.

In the modern world, this "cultural hypnosis" extends to experiencing a world in which scientific "laws" are always obeyed, whereas in other cultures "violations" of these laws may be relatively commonplace. For example, the phenomenon of achieving such a state of consciousness that one can walk with impunity barefoot, slowly, over a bed of hot coals, is one which is regularly observable in a variety of "primitive" societies; in modern Western society there used to be general skepticism about tales of witnessing this phenomenon, although in recent years some "power of the mind" workshops offer to teach the ability. In some societies remote perception (e.g., of the whereabouts of cattle that may have strayed out of sight) is regularly employed. In modern Western society the phenomenon was generally assumed to be "physically impossible"; however, in recent years there has been research with positive results, on both sides of the Iron Curtain, as the phenomenon turns out be useful for military intelligence purposes. Reports from the Indian subcontinent of extraordinary capabilities of yogis to control bodily processes with their minds were treated with considerable skepticism until the advent of biofeedback training showed that we all have far more potential ability than had been recognized to control brain waves, blood flow, heart beat, and the like.

It is humbling indeed for the typical Western scientist to realize the extent to which this sort of "cultural hypnosis" has shaped the scientific world view. Bias inevitably creeps into the construction of any knowledge system, because research into topics of interest receives more support and attention than research into matters of less concern. Thus it is natural enough that Western society, characterized by keen interest in manipulating and exploiting the physical environment, would emphasize and support research that appears to hold promise of generating useful technologies, that is, knowledge tested by its usefulness in prediction and control. But we are speaking of something more subtle than the emphasis on exploring the measurable world and the relative neglect of the world of inner experience. If observations are affected by expectation and suggestion, and if all scientists in Western society tend to have the same sorts of expectations and are vulnerable to the same sorts of suggestions, then how can we be sure how much science reflects reality and how much it is colored by this inner programming?

The question is neither idle nor polemical. It is characteristic of hypnosis that the person will rationalize as necessary to preserve the apparent integrity of what is experienced as "real"; obedience to a post-hypnotic suggestion, for example, will bring forth a full rationalization why that was a most reasonable thing to do. Thus if the possibility is conceded to be present, in any attempt to construct the "second kind of science," that we might be fooled by the pervasive Western "cultural hypnosis," the reality testing of our science appears to be a more subtle matter than might have been previously appreciated. For it follows that evaluation of scientific method through rational philosophy is not guaranteed to uncover biases and "blind spots"; these are most likely to be detected from a vantage point outside this culture, which puts such store by analytical, rational philosophy.

There has long been an esoteric tradition throughout Western (and Eastern) history, of the potentiality for the individual to become "dehypnotized" or "enlightened." A few quotations will serve as reminders:

> *In the ordinary waking state . . . man does not see the real world. . . . He lives in hypnotic sleep. . . . "To awaken" for man means to be "dehypnotized."*
>
> P. D. Ouspensky
> *In Search of the Miraculous*

Humanity is asleep, concerned only with what is useless, living in a wrong world.

Sanai of Afghanistan, A.D. 1130

Our conscious thinking has all the characteristics of a dream. . . . The representation that it gives us of the world is illusory. . . . In (a higher awareness state) consciousness is awakened in a way which is no longer exclusive or attached. . . . It is liberated from usual hypnosis.

Hubert Benoit, *Let Go! Theory and Practice of Detachment According to Zen*

We do not want to attain truth. We do not want anyone to break our dream.

Swami Vivekananda

Those who would contribute to a complementary science should pay heed to the implicit advice here, namely that they will inevitably encounter their own cultural hypnosis, and will be changed in the process.

A "Second Copernican Revolution"?

It is not true, of course, that the exploration of inner, subjective experience has been totally neglected. Data contributive to a "complementary science" are available from many sources: experimental psychology, cultural anthropology, psychotherapy, comparative religion, and parapsychology, to name a few. As already implied, the esoteric forms of the spiritual traditions of East and West provide a particularly rich source of clues; so also do the spiritual guidebooks emanating from the various monastic traditions, since these were of course the equivalent of research laboratories for inner experiment.

While these leads must be followed up and validated in some more public way before they add up to something one might begin to call a science, nevertheless the dim outlines of a paradigm which would have seemed quite alien to most scientists even a decade ago are beginning to emerge. It is not a pattern of concepts in conflict with those of conventional science, but one that seems to stand in a *complementary* relationship with it in the sense made popular much earlier in this century by the physicist Neils Bohr—namely, that reality is too rich to be comprehended in a single conceptual framework, so that seemingly conflicting metaphors (e.g., wave and particle representations of light) may instead be complementary, representing different observable or experiencable aspects of reality. Two specific contrasts with more conventional scientific views will help illustrate this point.

The first can best be conveyed through an analogy with dreaming. While one is in a dream, it all seems real enough. In the dream, various events take place, and there appear to be causal relationships among them. But then the dreamer wakes up and observes that the causality situation is really quite different from what it seemed to be in the dream. In the waking state it is obvious that "I, the dreamer" caused the events and the relationships among them. Now consider "ordinary reality." There are various kinds of events and apparent causal relationships between them. Some of these causal relationships are so dependable that we discover "scientific laws" to describe them. But once in a while someone "wakes up" from this world and observes that it is like a collective dream. In this moment of enlightenment, the person sees that "I/we, the dreamer" creates the events and the interrelationships; the nature of causality is totally different than appeared to be the case in the ordinary state of consciousness. This view of causality may seem more characteristic of the Eastern religious philosophies, but it does not in most instances contradict the usual scientific view.

The second contrast relates to evolution. The conventional scientific picture of the world, from a hypothetical "big bang" to the present, is a story of the material evolution of the stars and planets, of single-cell life, of plants and animals and eventually humans, with the culminating characteristic of man—self-consciousness—somehow evanescing from the physicochemical processes of the brain. However, when the complementary view is put together from the reports of inner exploration through various means and in various societies, what seems to emerge is a picture in which consciousness was here first! Consciousness appears in this complementary view to pervade all of the universe and to have pulled material evolution in certain preferred directions, most specifically the direction of increasing self-awareness. Again, the materialist view and this transcendentalist view may seem contradictory; more precisely, they are complementary.

It is interesting to note that the popular culture, particularly in the United States, has adapted to a view similar to that implied here to a remarkable extent in the past few years, as indicated, for example, by the kinds of books people buy (and presumably read). It appears that the cultural shift is in advance of new scientific understandings, which represents quite a contrast to the usual sequence throughout the history of science, where scientific discoveries were made first, only eventually affecting people's beliefs and values. (Remember, however, that in the Scientific Revolution itself, the cultural mavericks led the way and the church authorities came along later, and reluctantly.)

We recall how thoroughly society was affected in the end by the Copernican hypothesis. Yet it is easy to imagine that standing in the early seventeenth century one might well have argued that these speculations would be of interest to a few astronomers and philosophers, but surely of doubtful importance to the "man in the street"! In a similar way it may seem that the indications we have reviewed of a growing sense of need for an expanded scientific framework would have no obviously predictable impacts on the economy, national security policies, or technological development. Yet when we consider that once again basic concepts of the nature of reality, long taken for granted, are being challenged, it becomes apparent how profound a societal transformation may be foreshadowed by this "second Copernican revolution." Whereas the first Copernican revolution transformed our conceptions of outer space, the second transforms our view of "inner space," bringing with it a reconceptualization of ultimate human motivation and universal values, as well as a new consensus on the nature of the good society. As the first brought us the marvels of modern science and technology, along with toxic chemical and radioactive waste problems, acid rain, and the potential for nuclear holocaust, so the second may bring us the insight to achieve sustainable peace, common security, and a viable global future.

References

James, W. 1943. *Varieties of Religious Experience*. New York: Modern Library.

Sperry, R. 1981. Changing priorities. *Annual Review of Neurosciences*, vol. IV.

Wilber, K. 1983. *Eye to Eye: The Quest for the New Paradigm*. New York: Anchor/Doubleday.

Discussion

Willis W. Harman

Gilbert Durand: The unity of your paper interested me. What do you mean exactly by the concept of consciousness? I will be more specific: Do you mean by that the conscious as opposed to the unconscious, or are you bringing in something more ontological, like what we generally call the mind? I would like to know exactly what you mean.

Willis Harman: I prefer to use the word consciousness instead of the word imaginary because if consciousness in English corresponds a bit to what the French academics here who read papers at this meeting call "imaginary," then the word imaginary in English does not have that same meaning. By consciousness at any rate I mean everything in our universe that does not partake of the so-called physical reality.

Gilbert Durand: Yes, in this case I think that corresponds to what we call the imaginary in French, that is, the general rubric of representation.

Willis Harman: Then we agree.

René Berger: It is true that many university people, when you speak to them in their universities, tend to say that science has nothing to do with consciousness. Of course that is not the case with the university people at our meeting here. But I think that our university people here are right. Actually, there is no way to distinguish, as you yourself do, however, between "hard science," which would be the true science, and "soft science," which would be the science of consciousness. It is evident that what you are discussing on the subject of consciousness and the way you approach the problem of analyzing conscious phenomena is just as scientific as the hard sciences. So I think there is no room for a complex about these hard sciences! You could also say that science begins with the analysis of consciousness, and that this hard science is finally what can make explicit, or complement, or help this analysis of consciousness, which is, naturally, essential for each one of us. Hard science is really a complement of the science of everyday life, and everyday life is consciousness.

Willis Harman: I think that one good definition of science is that it is the democratization of knowledge. Science must not remain the property of great priests. American universities should not leave studies of consciousness, lacking credibility, to be carried out in the basement of some far-removed building. We must also stop thinking that there can be no true scientific knowledge when we undertake the problems of consciousness. Consciousness plays, of course, *an equal part* in the scientific problem.

Jean Charon: I would like to say something about Willis Harman's paper. I think that he is right, and that it is oh so right, to insist that research on consciousness remains even today the poor relation of science, and that considerable efforts should be made to give research on consciousness a scientific place which corresponds a little better to the central place this problem has in what our contemporary knowledge should be. I cannot be too pessimistic, however, because I think we're getting there. We don't have to push too much, since it is going by itself. Science and its methods are changing, but as Albert Einstein remarked, science is no doubt a mill that grinds slowly. . . .

I would like to use again the example I gave in my own paper: science is like a visitor who looks at a crowd on Earth from a very high altitude. To study the crowd's behavior, this visitor will begin by studying its overall movements, the most visible ones and also the ones that are repeated; and he will derive from that, no doubt, great statistical laws relative to its behavior. Then, as he analyzes the crowd's behavior more deeply, he will at some point be obliged to study the behavior of the individuals who compose this crowd. And then, of course, he will no longer find statistical phenomena, and he will also see non-repeating phenomena.

Physics has developed more or less like this, especially since Newton. First it looked for the most visible experimental facts, which allowed for the classifying of the repeating phenomena into laws. Then, starting with Einstein, analysis reached such a profound level and the phenomena became so complex, that as Einstein remarked and I mention again, the experiment would no longer provide the major criterion for establishing laws, but rather great principles of coherence, principles *of logic* allowing phenomena to be examined, or at least allowing us to foretell their behavior even before discovering experimental facts. Theory preceded experience.

Today we note the impact between the mental and the material, between consciousness and matter, and we are about to discover the realm of non-repetitive phenomena, taking into consideration the diversity of states of consciousness, not only of the observer's states of consciousness, but also the states of consciousness of the beings we observe. And now you know, because with my complex relativity I have become its chief advocate, that we can now represent these states of consciousness even down to the elementary particles of physics themselves. We are thus entering a new physics and a new science, where phenomena of consciousness fall within its direct field of investigation. Without the slightest doubt, and I agree 100 percent with Willis Harman, this will force us to expand the methods of science, which had been conventional until now. I would even go so far as to say that the true scientists, even if they do not admit it openly yet, know perfectly well that science today has opened itself up, and will continue to open up irreversably, to research on states of consciousness and on what may be called the structure of the mental and its interaction with matter. After all, our present gathering on "Spirit and Science" takes note of this evolutionary direction of science.

Thanks again to Willis Harman who, as you know, directs the Center for Research on Consciousness in San Francisco. Thank you, Willis, for your contribution to the scientific evolution toward research on consciousness.

7.
In Search of the Lost Reality

Jean Lerède

Reality is inaccessible to the thinker of the intellectual and rational type.

Pavlov

A Question Poorly Stated Because Poorly Experienced

Does the material world, the external world, exist independently of the representations of the human mind? Is there existence *in itself*? Is it *real*? Or does that external world, on the contrary, have no other reality than the ideas of "imaginary" representations we make of it? This is an old debate between the partisans of materialism and the advocates of idealism, the eternal question to which Buddha apparently judged it better to respond with a "noble silence." Montaigne later echoed him (if we may put it that way) when he wrote, "Of what use are these keen points of philosophy that exceed our understanding and our knowledge?" Both no doubt felt that the question, posed as an intellectual problem to be solved, was in fact of no interest because the answer, supposing there were one, would be of no use for the happiness of human beings.

Of no interest? Perhaps. But not without consequences. For the answer given to that question people have had their throats cut, been burned, drowned, guillotined, and faced firing squads with ardor down the centuries, and nothing would indicate that such ardor was beginning to decline in our times if there were not a certain number of converging indications that currently lead one to believe that we are perhaps on the point of having done with a question which, if not useless, is at least poorly stated, and poorly stated because poorly experienced.

Could not the question be formulated in the following way: "Is it possible to live daily, practically, concretely, our relationship as human beings with *both* the material, outer world *and* our inner world (the world of representation, the world of the imaginary in the broadest sense of the word) in such a way that the pressure of the "outer" on the "inner" is experienced not with pain and anguish, as is now the case, but in a balanced and happy fashion? Could the reality of *reality* be made up of the conjunction of these two opposites and their complementarity, not in a theoretical, analytical, and abstract sense, but in terms of the most deeply lived experience of the human psyche itself? Would the psyche not find a cure for its sufferings and tortured self-questioning in the dual reflection of matter and spirit experienced existentially, complementarily, in its most intimate depths? A double iridescence, not static but dynamic and evolving, an equilibrium in the process of growing and constant unveiling, which would imply in the last analysis the slow and patient progress of individuals and societies towards a reality that is *one* and not dual.

This reality would be concretely apprehended and lived by the human being as an experience constantly broadening and deepening, in the stead of the eternal, sterile, and bloody metaphysical quarrels on the respective natures and the relation between matter and spirit, a question truly insoluble as long as it continues to be posed in the terms of verbalized and intellectual thinking, as it generally has been till now.

In other words, can one conceive a peaceful, *unified* coexistence in actual life between the world of matter and the world of spirit that would make but *one* world, instead of the conflict-ridden, lacerating coexistence a human being currently experiences? A *systemic* and harmonious co-existence that would acknowledge the human being to be *poised* between matter and spirit? Is this possible and how?

No doubt our remote ancestors of the Upper Paleolithic era some fifteen or twenty thousand years ago did not formulate the question quite like that. But it is infinitely probable that the question had already presented itself to them, simply because they, too, were human. And it is also possible that to this question, this vital tension which already was threatening their happy adaptation to life, they had already found an answer, not intellectual, but experiential, and this was *initiation.*

The Imaginary, Royal Road to Lost Reality: Mythical Symbolism

The history of religions and ethnology teach us that one of the most significant and important psycho-spiritual phenomena in the history of humanity, and until now among the least studied in psychology, is that of initiation, which under various names has left a deep imprint on the course of human existence for thousands of years, very likely since the Paleolithic era, if not even earlier.

"By initiation," writes the great historian of religions, Mircea Eliade, "is meant a set of rites and oral teachings that aims at the radical modification of the religious and social status of the candidate for initiation. . . . Initiation is equivalent to an ontological mutation of the existential condition. . . . At the end of his ordeals, the neophyte leads an entirely different existence than he did prior to the initiation; he has become another person. . . . It is initiation that confers on men their human status; prior to the initiation, one does not yet fully participate in the human condition. . . . Thus, initiation constitutes a decisive experience in the life of every individual in pre-modern societies; it is a fundamental existential experience, because, thanks to it, man becomes capable of assuming fully his mode of being."[1]

"Most initiation ordeals," Eliade continues, "imply more or less transparently a ritual *death* followed by a resurrection and a *re-birth.* The central moment in any initiation is represented by the ceremony that symbolizes the death of the initiate and his return among the living. But he returns to life a new man, assuming a new mode of being. . . . This new life is conceived as the true human existence, for it is open to the values of the spirit. . . . The initiatory death is indispensable to the 'commencement' of spiritual life. The function (of this initiatory death) must be understood in relation to what it prepares for: birth to a higher mode of being."[2]

Initiation puts the human being in communication with the most intimate regions of the unconscious. It aims at freeing the individual from infantile submission to parental authority or the authority of social custom. Even more profoundly, initiation aims at releasing the human being from the tyranny of sensation and neurological subservience to the external world. Initiation is a form of shock treatment that is intended to make of the initiate a free and autonomous being, freed from the *hypnotic state* in which the fascination of the material world, the external world, plunges him, to borrow a striking expression from Professor Izutsu, now director of the Department of Philosophy at Keio University in Tokyo.

A few distinctions are appropriate at this point. The history of religions teaches us that there are in fact three kinds of initiation. The kind of initiation known as "puberty" or "tribal" initiation is obligatory for all the young people of the same human group. It marks the passage from childhood or adolescence to adulthood. It often includes the (symbolic) sacrifice or murder of the real father and mother, thus witnessing that the initiate is freed from parental guardianship. The initiation introduces the neophyte into both the human community and the adult world of spiritual values.

The second kind of initiation involves adults only, and represents the only true initiation, because it assumes the nature of a personal choice. It is practiced individually or in limited groups. It introduces the neophyte into a closed society or fraternity of a more or less secret or esoteric nature, whose goal is the initial awakening and the further spiritual progress of its members.

A third kind of initiation satisfies an inner call that is even more specific than for the initiation just discussed, and in fact constitutes only a variation of it. The differences are of degree and not of essence. Intimate personal experience here plays a predominant role. The neophyte aspires to a more decisive initial experience and a more intense psycho-religious life than is available to the other members of the community. In primitive religions, such an initiation marks the consecration of the *medicine-man* or shaman.

These three initiations, which can actually be reduced to two—one for adolescents, the other for adults—have been practiced in all archaic societies. Both rely to varying degrees on one and the same *reality*, considered essential in the psycho-biological development of any human being: the encounter with the "sacred," with a world of values absolutely other than those of the external, the profane world. Such an encounter also and more fundamentally establishes *a mode of psychological functioning radically different from that of "natural" man*, who is not yet initiated, not yet grafted. The men of those distant times believed that without such grafting, the tree of human nature could not bear good fruit; it would produce only thorns and wild fruits.

The encounter with a higher mode of being (that is, better adapted to life, more "realistic") could not come about in archaic societies without the intervention of a specific psychological process: *initiatory death followed by re-birth*. It was the task of myth and rite to bring about existential realization through the psycho-spiritual experiences called forth during initiation.

We should clarify what we mean by myth and rite from the perspective we are now assuming, which is specifically concerned with emphasizing the essential role played by the "imaginary" in the search for lost reality. Mythological symbolism —Egyptian, Chaldean, Uralo-altaic, Indian, Greek, Judeo-Christian, American Indian, etc.—is of a psycho-biological, ethico-genetic order. But the language of myths is an enigmatic language whose significance is not immediately apparent. Humankind has progressively lost the immediate, neurological sense of myths and symbols. Our symbolic function has atrophied to the point of disappearing. For man in modern societies, symbolic language is a "forgotten language," as Fromm puts it.

Indeed, it is helpful to consider the language of myths and symbols as a code language that has a literal sense and a hidden sense. There is an underlying, extremely precise coherence, logic, and significance in mythical tales that only the gradual elucidation of their hidden meaning can bring to light. The coherence, logic, and "system of laws" that govern myths are radically different from those of discursive reason, but they are just as strict. That coherence and that logic testify to the existence in the human race of a kind of thinking, a mode of being, a manner of apprehending the real, and in addition a kind of "law" that are absolutely different from those to which we are currently accustomed.

Composed of those elementary units consisting of symbols assembled in a given order, myths in the broad sense of the word are products, if not of the collective unconscious, as Jung maintains, at least of the "deep imagination" within the individual unconscious. The symbolic images are *the* language of the unconscious as it is expressed in art and especially in myths and dreams. Myths are synthetic and condensed key-images, images saturated with significance, whether or not organized into tales, where the cosmogony and adventures, or simply the images, of the gods and heros as well as of those deified ancestors or symbolic animals illustrate and unveil the most intimate mechanisms of the human psyche on the unconscious level. The myths reveal the most profound aspects of *inner reality*, which rational language is powerless to translate. As Diel writes, "Myth is an intimate psychology that is expressed with the aid of symbolic terminology."[3] Says Eliade, "Myths answer a need and fulfill a function: that of baring the most secret ways of being."[4]

"Myth expresses the world as perceived and experienced by the subject, not according to his critical reason and on the conscious level, but according to his whole affective and representational psychic system, principally on the unconscious level," writes J. Chavalier. "Like a scouting party sent ahead into unknown territory, myth scrutinizes and tends to express the direction of the spiritual adventure of men advancing through space-time."[5]

Myths signal the vitally sensed solutions to the problems the human individual faces in adapting to his inner experience, his relation to his fellow humans, and the environment. Hunger and thirst, the sexual instinct, suffering and death, love and joy, good and evil, thought and action, the conquest of liberty, the direction of life, the relation between matter and spirit, the intimate nature of reality: to the instincts, needs, questions and fundamental problems of human beings of all eras myths offer the answer, expressed in the only language that can authentically be the answer for all, on the level of the unconscious. Myth links the individual to the collective and the universal because myths and symbols are *the language* of the universal and immutable through the ages and the successions of religions and cultures. All that changes is the form of the symbols and myths in the proliferation of imagery in which they are clothed and in the evolving biogenesis that continually works to make and re-make the synthetic and condensed expression of the unitary energy of the spirit.

Regarding the universality of myths, Gusdorf writes: "They recur from one end of the earth to the other throughout time and space. This diffusion is so general that it rules out any possibility of transmission simply by contact of one culture zone with another. We are forced to admit that there is a primordial spontaneity in the tale-making function that invents nearly the same stories everywhere there are human societies. . . . Thus, we find revealed an objective structure in mythical consciousness, itself the expression of a profound objectivity on the part of man in the world."[6] Durand calls myths the "anthropological structures of the imaginary," reflections in the very heart of the human psyche of the deepest laws of nature and life. Gusdorf goes on, "If discursive (and rational) validation is lacking (in the myth), this is not to say that the myth has no criteria. The very strength of its hold on man implies the proof of that. The truth of myth is attested to by the total impression of involvement it produces in us. We do not justify myth; rather, myth justifies us. It illuminates a situation, and suddenly we have a sense of being in a familiar place. The truth of myth reintegrates us into the whole, by virtue of a function of ontological recognition."[7]

Myths and symbolic images are not just "representations." They do not merely illustrate, unveil, and give signs. They are also programmed *media* and extraordinarily powerful transformers of psychic energy. To them can be applied the definition that R. Cahen gives of an archetype: "A *cybernetic program* that imposes a collectively inherited solution on a perennial relational problem."[8] Myths and symbols refer to themselves, symbolically, as "the Word of God." They have healing power over the sick psyche, which they calm and enable to go through the steps required for a return to equilibrium. Myths simultaneously tear us down and build us up; they put us back together. Myths restore our sense of personal responsibility towards life. They also unmask us, helping to break down our pride, our inner lies and our tendency to accuse others, all elements of an "unrealism" that prevents us from seeing beings, things, and ourselves as they are and as we are. All these elements interfere with the essential function of the imaginary, which is, paradoxically, to give us, or restore to us, the sense of the real. Myths tell us in their own language that each person is responsible for the state of his own psyche and that we are to a great extent the makers of our sufferings and inner distress, the artisans of our own maladjustment to the real. Myths, as Jung very rightly points out, are in reality psycho-therapeutic assemblages of monumental proportions, for which religions have served as a framework.

As Cassirer writes, "Myth is objective because it allows consciousness to escape and set itself against mere receptivity to sense impressions."[9] Myth and myth alone allows us to filter, transform, and organize such impressions and to do so at the unconscious level where myth operates, exercising its function of "regulating the ontological equilibrium of man," to use Gusdorf's expression.[10] By the changes it brings about in the psychological organization of the individual on the unconscious level, myth "vaccinates" against the excessive impact, the physiologically dangerous and psychologically deadly pressure that the external world exercises on him through the nervous system. In other words, myth, when it is set in the very heart of the human psyche, there re-establishes the broken equilibrium between extroversion and introversion. Myth, product of the "imaginary," reunifies the human psyche. It restores to it an undistorted, realistic view of the real.

Manifestations of an organizing and reshaping dynamism, as G. Durand defines them, myths also have the power to waken the latent and potential resources of the human psyche. Here we have another fundamental aspect of reality: the qualities and potentialities of the human being that are not yet realized, not yet observable. "Mythology," writes Gusdorf, "provides an inventory of human possibilities."[11] Using an enigmatic language akin to that of dreams, myths, like dreams, are actually talking about the healthy or unhealthy functioning of the psyche, in terms of the evolutionary adaptation of the human species. Myths have the capacity to stimulate psychologically not only the process of adaptation, but also creativity. They unleash and foster in man the unconscious dialectical mechanisms that tend to bring about a higher mode of being. Myths and symbols facilitate the slow unfolding in the human race of that psychic instance in the process of evolutionary formation that I propose to follow Diel in calling the "supra-conscious," which has the power, among other aspects, to reconcile the conscious and the unconscious in man, creating a relationship of functional synergy between them.[12, 13] Myths and symbols, in short, have a five-fold function: pedagogical, therapeutic, organizing, regulatory and evolutionary.

There are, however, good and bad uses of symbols and myths. Myth is a dangerous and delicate product to handle. If a myth is taken for spatio-temporal reality and is apprehended intellectually, as was the case with the historical religions, myth becomes superstition and dogmatism. Far from liberating the spirit, it burdens and deforms it. Far from driving out fear and putting an end to isolation and anguish, it reinforces and aggravates them. Far from uniting men, it divides them and pits them one against another. A symbol that has grown cold, to borrow an expression from Hegel, a myth that has become an intellectual concept, wreaks havoc on the individual and collective psychic system. The latter kills the spirit. It makes it an abstract entity separate from matter and in opposition to matter. Myth more or less unconsciously rationalized leads to dissociation, the splitting of the human psyche, henceforth incapable of thinking and generally of living life in terms other than *oppositions*, because opposition is the law of the rational intellect, the law of Aristotelian logic. A law can be fertile and *realistic* within its own sphere, which is the spatio-temporal domain, but a law has fatal effects when it claims to exercise its empire over the inner psyche and penetrates—presumptuously and *unrealistically*—into the deep, non spatio-temporal places of the human psyche.

Myth is actually a category of sentiment—of sentiment and not of sentimentality—a category of emotion, of the imaginary and of the unconscious. Even more specifically, images and myths, profound expressions of *superconscious* intuition, act directly on the sensibility and on the conscious and unconscious imagination.

But to assume life, myths and symbols require the temporary and freely granted silence of reason and the theoretical and practical intellect. On this point, as on others just as fundamental, I do not share the reductionist and rationalist views of Diel, in spite of the great value I personally attach to his understanding of mythology, which in my opinion leaves both the work of Freud and that of Jung far behind.[14] Along with Durand, I believe that the obscuring of what Bachelard calls "the symbolic function" is a fundamental sickness that has stricken the human race for a long time and never more than at present. It is a mutilation with dramatic consequences. As Durand writes, "What is needed is to demystify the symbol and at the same time to remythify it."[15] What is needed is to rediscover in ourselves the use of our *lost intuitive understanding*, obscured by the unhealthy proliferation of merely intellectual understanding. Myths and symbols can restore that understanding, providing that we re-learn their proper use.

I agree completely with Diel's insistence on the rational and psychological elucidation and decipherment of symbols. In the present conditions of modern humanity, myth must be translated into conceptual language to be understood and to become efficacious. As H. Corbin also thinks, even though his standpoint is very different from Diel's, the hermeneutical study of the various symbolic traditions is an urgent task of our times, probably the only way of calling forth the new psychobiological values—new and at the same time eternal—that present-day humanity so desperately lacks. But I cannot follow Diel when he advocates abandoning symbols in favor of their rationalized translation, though this does not imply my attachment in principle to images.

Understanding a symbol rationally is not the same as making it disappear. On the contrary, the mythical image is revivified by rational translation, which, by expressing the symbols in modern psychological language, simply makes understanding easier and less hazardous and accords to reason the legitimate role it should continue to play in the whole process of symbolic assimilation. That same mythical image, however, or more broadly speaking that same mythical language, remains more necessary than ever.

As products of the superconscious imagination, myths and symbols are dynamogenic at the deepest level of the psyche; their rationalized translation is not. It is semantics dried up into semiology, as Durand would say. Translating a myth, moreover, is in reality nothing but substituting one symbolism for another. The myth translated becomes conceptual symbolism. It is a useful, even indispensable symbolism, from two points of view. First, it keeps the myth itself from being taken superstitiously for a reality of the spatio-temporal kind. Second, our fundamentally rationalized minds probably need such a translation in order to waken their sleeping mythic sensibility, as a kind of bridge between the forgotten world of intuitive and symbolic thought and the rational world of concepts in which our thinking habitually moves. But this conceptual symbolism, which results from translating the myth into psychological language, replaces a rich symbolism with a poor one. It is a symbolism without images, with wings clipped. In short, it would be a dry and abstract symbolism, bones without flesh, a dead symbolism, if it misunderstood its irreducible nature as symbol. Perhaps it would be more appropriate to speak in this context of *symbolizing concepts*, whose value and efficacy are the greater because a more or less submerged spatial or temporal image subsists in them. Examples are superconsciousness, and subconsciousness. These symbolizing concepts can be a powerful aid in the restoration of the symbolic function in a human individual, which once established will institute a mode of thought radically different from conceptual thought.

Along with Corbin, I believe that mythical symbolism is the best and, to tell the truth, the only possible expression of certain profound psychic phenomena that never rise to the level of consciousness. Again with Corbin, I also believe that there are "degrees of penetration of the real into reality," and that beyond the spatio-temporal world there exist other realities, though on a deeper level, that are just as "real," the realities of an "intermediary world," which is none other than the world of superconscious realities in the human individual.[16] Corbin strongly maintains that there is a spiritual realism, that of an imaginary, or better, an "imaginal" world, to which we have access through myths and symbols. But the fundamental error is to confuse that reality with the reality of the world of space and time in which we live day by day, even though, paradoxically, the two realities are bound in an indissoluble unity within the human psyche. Another error, no less fundamental, would be to confuse the affirmation of that realism with the immemorial, and to my mind immemorially false, positions of philosophical idealism.

Moreover, I believe with Eliade, that symbolic images can never be exhausted by the various "translations" that may be proposed. Interpretation forestalls the abuses of the imagination; it saves the imagination from its temptations. It wards off, as we have said, the danger of superstition, that infantile —or just as likely senile—disease of the symbol that comes from mistaking it for spatio-temporal reality, the evil that has gnawed at all historical religions. But interpretation is no substitute for the mythical images themselves, and if we insist on a definitive and overly rigid formulation, the rational translation will purely and simply destroy the symbol it claims to clarify.

Eliade writes in this connection, "If the mind uses images to seize the ultimate reality of things, it is because that reality is manifested in a paradoxical fashion and therefore cannot be expressed by concepts. . . . It is the power and task of images to *show* whatever remains refractory to concept. This explains the misfortune and downfall of the man who 'lacks imagination'; he is cut off from the deeper reality of life and his own soul . . . Thus, it is the image as such as a cluster of meaning, that is true, and not *one only of its meanings or one only of its many planes of reference.* To translate an image into concrete terminology by reducing it to only one of its planes of reference is worse than mutilation, it is tantamount to destroying it, nullifying it as an instrument of knowledge."[17] Jung writes, "As soon as one tries to get at the 'essential nature' of the image by abstraction, it becomes nebulous and goes up in smoke. To understand its living function, one must allow it to subsist in its complexity as a living being; one must not try to analyze it scientifically on the basis of the anatomy of its corpse, or historically on the basis of the archaeology of its ruins."[18] And Gusdorf writes, "The takeover of the myth by the intelligence, its reflective transcription, allows the essential to escape. . . . Mythical consciousness gives rise to engaged thinking, which one cannot disengage without falsifying it."[19]

One can no more "explain" a myth "rationally" than one can explain a Bach chorale or a Beethoven symphony, even though musical analysis and criticism are capable of contributing powerfully to the understanding of a work and increasing the awareness of those who perform or listen to it. As

Corbin emphasizes, "A symbol . . . is never explained once and for all, but must continually be redeciphered, just as a musical score is never deciphered once and for all, but continually calls for a new performance."[20] "A symbolic image," Durand writes, "must be ceaselessly reexperienced, somewhat as a piece of music or a protagonist in a play requires an 'interpreter.' "[21]

A myth, as Eliade says, is a cluster of meanings, and they are on such different levels that they often appear contradictory. In reality, they are contradictory only to rational thought, frozen in its fundamental incapacity to apprehend the union of opposites that the mythical image specifically expresses. The myth is inexhaustible. Analyzed, it immediately invites the analyst to explore further. And the analyst will rather tire of conceiving than the myth will of supplying, we might say in paraphrase of Pascal. Our comprehension of the myth, in fact, will be on our own personal level of development and understanding. As D. Cousineau justly points out, "The human psyche functions precisely according to the dynamic and paradoxical model that characterizes the mythical world."[22]

Allegory is one thing, symbol another. It is essential not to confuse them. Allegory—the traditional female figure representing Justice, for example, bearing her allegorical attributes such as the scale and the sword—is "simple concrete translation of an idea difficult to grasp or express simply."[23] Unlike symbol, allegory in no way refers us to a different order, another world. Allegory is just ease and simplification of language, whereas symbol is "a representation that makes a secret meaning *appear*; it is the epiphany of a mystery."[24] "Once the translation is made," writes P. Ricoeur, "one can drop the allegory, henceforth useless."[25] Just the opposite is the case where a mythical symbol is concerned. The latter, however, is degraded to the level of allegory when it is emptied of meaning and ceases to refer to another world.

In short, I believe along with Eliade that one must guard against mutilating the imagination, the inexhaustible source of inner riches and creativity, and should instead "reawaken the images."[26] To which statement I would simply append the following: after having elucidated the meaning at the level where it is possible to do so.

Mythical images, in their recovered efficacy, demand that reason be silent. But putting the intellectual functions on the back burner by no means implies repudiation; reason retains its rights. It is merely suspended for a time, even if for just a fraction of a second. Perhaps it would be more accurate to say that reason ceases to be active, is "disconnected." By accepting this "disconnection" of reason, reason itself is conscious of acting in the most reasonable way possible.

When reason sleeps, or, to put it more accurately, when the intellect is calm and keeps still while the conscious mind nevertheless remains on watch, when the human psyche thus reaches a paradoxical state of profound, alert relaxation, an "attitude" of *active passivity* (an expression Eliade makes use of), a state of receptivity, of *inner silence*, it is only then, in my opinion, that myths can freely penetrate the unconscious and fulfill their mission of wakening and stimulating and thus show themselves fully operative without declining into sterile dogmatism or superstition.

The same holds for rite, which is nothing but one of myth's modes of action, merely a concrete medium, a sign language, for myth, but one of prime importance in archaic societies. Rite exists only thanks to the myth that it in turn renders more fully experienced, more present, more effective. Rite is myth mimed, felt and fixed in the body and the sentiments, while reason maintains silence. Rite is the visual, gestural, aural and perceivable *mise en scène* of the condensed psychodrama that is myth. Rite is "myth in action," as Van der Leeuw says.[27]

Prehistoric Man, Two-Level Man

Let us go back to the prehistoric era we briefly evoked earlier. The scope of this article does not permit me to present arguments here for the hypothesis that the prehistoric caves (with their frequent animal depictions) were sites of initiation where the existential status of the individual and his intimate psychological functioning were radically altered.[28] I do believe that the bulls of Lascaux or the bisons of Altamira, hidden in the depths of the silence and night of the caves fifteen thousand or more years ago, served a specific initiatory and therapeutic function, using the suggestive power of such a decor to illustrate and better yet *engrave* on the minds of the candidates for initiation the old primordial and dialectic myth of the human condition, the key since prehistoric times to all the developmental metamorphoses of the deep unconscious: *one must die in order to be reborn*, one must die to attain a certain kind of consciousness, formed, or rather deformed, merely by activity directed towards the external world, in itself a factor in the rationalization of the mind that engages in it. Such a death, at the cost of a fundamental shakeup of the conscious and unconscious structures of the inner psyche, brings about in the individual a real "resurrection," a mutation, a progressive and dialectical reconstruction of his psychological and neurological being that makes him a *man of two levels*, henceforth capable of functioning more and more harmoniously in the two keys of profound intuition and practical reason, capable of apprehending reality in its two aspects: inner and outer, "imaginary" and material, intuitive and rational.[29]

Study of the roots of Indo-European (fossil language of our distant prehistoric ancestors and mother language of almost all the modern European and Southern Asian languages) provides additional support for this hypothesis; most of those roots, as it happens, strikingly reflect the old primordial myth of inner death followed by a second birth and a radically new type of consciousness and thought.

Might the "man of the caves, two-level man" have possibly known greater balance and a sense of reality that would later be lost and for which we, men of the twentieth century, might well envy him? Can we really believe it?

Let us consider a February 7, 1979, *Le Monde* interview between journalist Jean Contrucci and Escalon de Fonton, scholar of prehistory, director of Research at the National Scientific Research Center and director of Prehistoric Antiquities for the region of Provence—Alps—Cote d'Azur. Fonton's statements shed a certain light on what we have just been saying about our distant ancestor, the two-level man. "Prehistoric man lived in peace with his kind. . . . The causes that lead to war did not yet exist. . . . The prehistoric hunter almost never turned his axe or hunting spear against his fellow man. He saved his energies for the pursuit of game, to which he owed his survival."

It stayed that way for some tens of thousands of years. Hunters are peaceable. In their habitations, we find thousands of weapons lying by the side of hearths, but the burial sites never reveal signs of any conflict. The only wounds found are those received in the hunt. There are no wounds of war. It is during the "dark ages" of the Neolithic, during the transition from the world of the hunters to agricultural and pastoral life, to settled dwelling and the idea of property, that the transition from a state of peace to a state of war unfolded. Only then did man become a wolf preying on other men.

In the masterly work that he has devoted to the Popol-Vuh, the sacred book of the Quiché-Maya of Mexico and Guatemala, Raphael Girard also states that during ancient prehistory, some fifteen to twenty thousand years ago, man lived without disputes or quarrels. War did not exist. Peaceful and good-natured, the men of those distant times did not even engage in hunting. They fed themselves on fruits, green vegetables, tubers, and wild corn. Monotheists, but conducting no worship or ceremony and hence lacking temples or priests, the Quiche-Maya of ancient prehistory were also unfamiliar with animism, magic, and totemism, which appeared only much later (cf. R. Girard, *Le Popol-Vuh*, Paris: Payot, 1972).

We find another significant analogy in the recent observations of the American ethnologist Kilton Stewart about the Senois. The Senois, a small tribe of the tropical forest of Malaysia, were living, at least until a few years ago, in conditions close to those of prehistoric man. According to Stewart, the Senois are unfamiliar with war, crime and class conflict. They live at a level of inner and interpersonal integration that is high and harmonious. In their case, too, it is striking to note the major role that the primordial initiation myth of inner transformation plays in that integration; one must "die" (in this existence) to be "reborn" (again in this existence).

But if all this should ultimately prove to be correct, two questions arise: *why* and *how* did humanity of the earliest times (presumably *happy*, i.e., realistically adapted to the human condition), lose that "blessed" state? Why and how did that "paradisiacal" state become a "paradise lost," as the myths known as myths "of the Fall," found in practically all archaic civilizations, have continued to proclaim for millenia in their symbolic language and under the most diverse forms?

How did we end up in our present state, where in spite of prodigious modern conquests in science and technology, humanity as a whole seems to us, as the twentieth century draws to its close, more anguished and *unhappy* than ever and threatened in its very survival? What has happened since distant prehistoric times?

Man After the "Fall"

Let us recall a few facts. The peoples of the Upper Paleolithic, twelve thousand years ago and more, lived in societies with economies and civilizations based on hunting, fishing, and gathering. These are the peoples who lived prior to the appearance of livestock breeding, agriculture, and private property. Beginning in the Mesolithic, throughout the entire Neolithic and down to the beginning of historical times, from 10,000 to 5,000 or 4,000 B.C., new types of economy, pastoral, agricultural, and even artisan, begin to take hold, probably as a result of a combination of climatic changes and technological developments. These new types of economy give rise to new structures of civilization, both material and social; with division of labor, specialization, and bartering of goods we already have the beginnings of agricultural capitalism, linked to private appropriation, and the appearance and general spread of the urge for individual ownership. About 6,000 B.C. we find the first dwellings clustered into villages, then into cities. Soon the first hierarchies based on religious, social, or military subjection were established. Rich and powerful cities begin to quarrel among themselves over material goods and political hegemony, whereas a nomadic way of life keeps primitive tribes from the critical danger of subjection to the same material goods. Excessive accumulation of riches by some arouses the envy of others, incites them to warlike aggression, and leads to collective armed conflicts. Kingdoms are organized, followed by empires, which require a central authority, administrators both economic and social, job specialization, hierarchical organization of public services, religion associated with the powers that be, in short, techniques of government and sophisticated organization to manipulate and dominate minds.

Well before then, the increasingly settled existence and plain accumulation associated with stock breeding, agriculture, and craft industry had led to the appearance of a new kind of thinking, more rationalized and static, and a corresponding decline, increasingly noticeable, in the mythical kind of thought and consciousness, of intuitive consciousness. The latter was no longer natural to man after the "Fall." But from what archaic societies have to teach us on this subject, it would seem that *initiation* nevertheless rather easily restored the individual to that state or at least partially revived it in him without excessive difficulty. As Gusdorf writes, "Primitive life [consecrated] the equilibrium once found between man and his environment in a lifestyle. The transcendent anchorage of myth consolidated the installation forever, once in place."[30]

Prehistory is the age of myth. *Homo religiosus* is the man of the societies that predate stock breeding and agriculture. He is prehistoric man. It is highly likely that prehistoric humanity found it easy to recapture the lost mythical thinking, intuitive thinking, and to live and move in it, an ease that disappeared little by little with the advent of more or less rationalized pastoral, agricultural and craft societies that were established after the close of the Upper Paleolithic.

"The transition from mythological thinking to rational thinking," writes Leroi-Gourhan, "was very gradual and fully synchronized with the development of urban concentration and metal-working."[31] The transition was furthered and accelerated by the appearance around 3,500 B.C. of the first forms of writing in Mesopotamia, followed around 1,500 B.C. by the first Phoenician consonant alphabets and around 750 B.C. by the Greek alphabets incorporating vowels. Writing was a major agent in the rationalization of thought.

What is writing? It is a means, required by the new economic and technological methods, relayed by language, itself a major instrument in the rationalization of thought, for retaining in the memory an exact account of the acquisitions made in the various spheres of human activity. "The technical efficiency (of writing) is in proportion to the elimination of the halo of associated images that characterizes archaic forms of writing," writes Leroi-Gourhan.[32]

The linear style of writing is undoubtedly a marvelous instrument for expressing and preserving thought. But what kind of thought? A kind of thought "more and more channelled into reasoning," Leroi-Gourhan goes on to say, "where the constriction of images (is the consequence) of the rigorous linearization of symbols (to which) writing tends . . . Letters make thought a penetrating line, long in reach, but thin as a wire. . . . Written language is designed to convey in economic fashion poor but precise ideas, whose linear adaptation assures their efficiency."[33]

In other words, the price of economic and technical precision and efficiency has been the impoverishment of human thought, a mutilation that implies, both in terms of inner developmental creativity and of expression and communication, thinking that is more and more preoccupied solely with the organization of the external world. This mutilation also implies that rational thinking substituted for what we have been calling the thinking of the two-level man in whom mythical, intuitive thought co-exists in balance with rational thought. To put it yet another way, man's tighter and tighter control over the material world is counterpoised by the tighter and tighter control of the material world over man and his thinking. The problem is older than modern times.

To mythical thinking corresponds a multi-dimensional form of graphic expression, of which the cave paintings of animals remain the clearest witness. Flying animals, rendered weightless, gallop in worlds without limits, outside of space and time. We find the same when we examine the language of prehistory, Indo-European, whose roots are essentially *polysignificative* and furthermore strongly reflect intuitive thinking, freed from the constraints of space and time. As Leroi-Gourhan observes, "Behind the symbolic assemblage of figures there must have existed an oral context with which the symbolic assemblage was coordinated and the values of which it reproduced spatially. . . . The greater part of the evolution of *homo sapiens* was accompanied by forms of thinking that have become strange to us, although they continue to underlie a major portion of our behavior. . . . Graphic expression restores to language the dimension of the inexpressible, the chance to multiply the dimensions of fact in immediately accessible visual symbols."[34]

The cave "mythograms" transcend space and time, another characteristic of mythical thought. They express free thought, radiating outward, unwritten and preferably non-verbal thought, moving with ease through a supple, multiple, non-fixed space-time constantly called into question. Such thought is also characteristic of the reborn individual, whose personality and consciousness, also multidimensional and *polycentric*, are reflected by it, and who avoids that "imprisonment of the world in the network of precise symbols" that Leroi-Gourhan speaks about.[35]

The new concept of space and time, rationalized, precise and efficient, was developed by the agricultural peoples to meet the new needs of their economy and social existence. But that same concept insidiously annexed a territory that was not its own, that of inner space and time, replacing freedom with constraint and introducing rigid structures forged by the practical and utilitarian intellect in a domain to which those structures were and should have remained foreign.

Agricultural societies were caught in the trap of space-time. The space was delimited, with finite surfaces and agricultural properties; hence, a finite inner space as well. Time was caught in a rhythm of labor, sowing, harvesting, and the changing round of the seasons; hence, inner time henceforth was enclosed and subject to structures imposing on it a pseudo-continuity foreign to its deepest nature.

The space-time trap that closed in on man in the societies that succeeded prehistory is also and increasingly the trap of history, of historicity. If prehistory is without history, it is not only because of the absence of written documents. It is also fundamentally because, as Gusdorf writes, "Prehistoric man does not bear witness to himself."[36] Part of the reason he did not invent writing was because he did not feel the need to set down his testimony to his times. That would constitute a security he did not need, an identification of self by self he did not have to make; myth sufficed for him; he lived in a daily absolute in which "there was nothing to hold onto of a reality always identical to itself."[37] Prehistoric mankind is ageless down the course of the millenia. As Gusdorf further emphasizes, "The historic mind-set is characteristic of the modern West. . . . There is no history of black Africa. . . . Nor is there a history of the Oceanic world, (nor) a history of the Chinese world or the Hindu world. . . . The sphere assigned to prehistory should be extended a great deal. . . . (It) encompasses the greater part of the history of humanity down to our own times.''[38]

The progressive reduction in free and creative *individual interpretation* attaching to mythical and graphic symbols; the decline of *sensory acuity*, and particularly the more specifically nonintellectual senses—taste, smell, touch—sources of knowledge in their own right and sources of apprehension of the real; the impoverishment of the *imagination*; a growing atrophy and sclerosis of spontaneous *memory*; a regressive change in the *will* severed from the natural and concentrated energy it drew from mythical sources and henceforth subject to the law of conscious and reasoned effort; less and less frequent recourse to nonrational, nonverbal and in particular sensory modes of expression and communication; and above all a fundamental anguish born of the collapse of myth in the psychological economy of the individual: these are some of the consequences of this mutation in the human mind, the loss of the mythical and intuitive thinking that had characterized the Upper Paleolithic thinking unlike any other because the specific environmental conditions that made it possible have disappeared for good.

We conclude from the foregoing pages that the thinking and consciousness of prehistoric man was probably exceptional and unique. That alone would justify the no less exceptional interest we have accorded to prehistory and the primordial myth on an issue that involves not only the structure of thought and the problem of its origins, but also the function of the real itself and our way of apprehending and experiencing reality in its two aspects: outer and inner. With regard to the origins of thinking, we cannot overemphasize the importance of the remark by Leroi-Gourhan, cited above, that the mythical thinking of prehistory has become strange to us, but continues to underlie a major portion of our behavior. Moreover, as A. Varagnac has shown, did not the traditional peasant cultures in Europe carry on the rites and customs of prehistoric times right up to the middle of the nineteenth century?[39]

The truth is that we still carry around in us the mythical kind of thinking, the thinking of prehistory. Dreams and their decipherment provide the most convincing proof. The key to explaining dreams is precisely the same as the key to interpreting myths, and it is impossible, in my opinion, to analyze dreams without applying the very same principles that enable us to understand mythical tales. Mythical thinking and dream thinking are two aspects of one and the same reality, buried deep in the psyche, and it is striking to encounter in the dreams of men and women of the twentieth century, alongside the obviously personal symbols, the ancient myths of all the religions of humanity. Myths and dreams use the same "imaginary" language to unveil the deepest psychological conflicts and to indicate to the individual how to restore his psychic system to healthy functioning and find the developmental energy needed to accomplish that end.[40]

Driven back by the imperialism of reason, mythical thinking has retreated into the unconscious regions of the psyche, where, being out of balance itself, it unbalances us. Could it be that the thinking of prehistoric man, which seems so distant to us, is with us much more than we can imagine? And might we have a vital interest, individually and collectively, in concerning ourselves with it?

A Lost Secret

One must die to be reborn, says the primordial myth of Paleolithic prehistory (and perhaps earlier), as we have said. We subsequently encounter the same myth in diverse forms, from the ancient Indian Vedas to the Gospel of Jesus.[41, 42] And in fundamental fashion, at that. But the rationalizing deformation of the human mind has brought it about—this is my own opinion, for which I take sole responsibility—that in all the religions of historic times the myth is found obscured, distorted, adulterated, rendered unrecognizable, because it is rationalized and institutionalized, and thus loses all reality and psychobiologically transformative power over human thinking and consciousness. The letter kills the spirit. With only the rarest of individual exceptions, the anticipated psychospiritual "mutation" has not taken place in humankind. One can see the same development everywhere: the brief resurgence of mythical thinking that appears in the beginnings of this or that new form of religion declines almost immediately into dogmatic rationalization, to a more or less pronounced degree. That rationalization gives rise to various forms of institutional and authoritarian religious orthodoxy, where superstitious belief in the letter of the mythical images and the efficacy of rituals, ceremonies, techniques, or works supplants the authentic and suggestive faith that the original message was trying to awaken. The evolutionary impetus, arising from life's mysterious depths, here and there throws off flashing sparks, but the sparks do not ignite the dry wood. Neither the New Testament, the Bhagavad-Gita, the Tao-te Ching, the Mayan Popol-Vuh, nor the Koran led to a mass reintegration of the human race into the lost "deep intuition." On the contrary, the rationalizing, and *masculinizing*, entropy that has affected the human race for thousands, perhaps tens of thousands, of years has become only more accentuated down the course of the centuries and has also accentuated the deepening split between the religion of the mass, a religion more and more rational and materialistic in spite of some appearances, and the religion of the adepts, a very restricted number, based on a personal, more or less profound, more or less authentic, initiation experience.

The prophetic views of Jesus on that subject could hardly be less optimistic. Passages abound in the Gospels (not to mention the book of Revelations) that bear witness to an extremely somber vision of the future of a humankind for which intuitive reintegration becomes increasingly difficult: "And the light shineth in darkness; and the darkness comprehended it not" (John 1:5–12). "For many are called, but few are chosen" (Matthew 22:14). "Straight is the gate, and narrow is the way, which leadeth unto life, and few there be that find it" (Matthew 7:14). "Men loved darkness rather than light. . ." (John 3:19). "When the Son of Man cometh, shall he find faith on the earth?" (Luke 18:8).

And yet other passages from the same New Testament text —the Sermon on the Mount, the Sermon at the Lakeside, the last chapters of Revelations which evoke the New Heavens and the New Earth—strike a more optimistic note. Those passages seem to foresee both an ultimately much *deeper* understanding of the mythical message itself and a far *broader* experiential realization of that message among the human race.

Undoubtedly it is the primitive Shi'ite tradition that has most clearly expressed, with respect to Islam, to be sure, but in terms that give it a much more general application, the idea that both the understanding and the realization of the original initiatory messages will be the task of the future, a distant future. The cycle of revelation (the "Seal of the Prophets"), which closes with the "sacred" text deemed to be final and to which nothing more can be added, is followed by the long, long cycle of the unveiling of that text and the progressive realization of the vital truths it contains. It is the cycle of *ta'wîl*, which means "taking something back to its source." It is the time of discovery of the hidden significance of the revelation. It is the time of initiation. It is the time of the slow historical maturing and working out in the inner man of the esoteric meaning of the original text. It is also what Shi'ite Islam calls the cycle of the *walâyat*, the cycle of apprenticeship in inner love and friendship among men. It is the time of the "hidden Imam," the Resurrector, the invisible and intimate Guide; it is the time of the inner Christ, the Savior and Paraclete of the Gospel tradition. But it is also the time of the Samadhi or the yogis of India, of the Buddhist Nirvana, the Chinese Tao, the Zen Satori; so many symbolic terms to designate, at various degrees of realization, the superconscious principle and its slow unveiling, its slow return to the human soul, as that soul frees itself from the "I."

The "I" is the "old consciousness," the "old man" of the mythical religions, who *must die* in order to be reborn; it is the "anatta" of Buddhism. The death of the "I" is the nonstruggling, nondesiring, nonsubstantiality of beings and things, the Void of the Tao, of Buddhism, of Zen.

The "I" is also the *hyper-rationalized* and intellectual (pseudointellectual) consciousness which has stifled intuition.[43] The "I" is the direct product of "objective" thinking, of "scientific" thinking, which renders man an object and treats him as such in the name of an "objectivity" valid in the physical sciences, but mutilating and thoroughly "unrealistic" when one attempts to apprehend the totality of the human being. It grasps only the outer, material, or social aspect and is unaware of the inner, "subjective" aspect, where the individual is revealed to be a "subject" (we do not mean to imply that the "objective" point of view applied to man cannot be valid, only that it is limited to the spatio-temporal realm). The *objective*, objectivized "I," *reifies* the individual, makes a thing of him, dehumanizes him, kills his sources of life and inner creativity, kills his adaptation to the real by reducing that function to the material world only and to the reason that reflects the material world in the human mind. "The real is the rational and the rational is the real," said Hegel, echoed not only by Marx and Lenin, but by the entire rational, "scientistic," and technical civilization of the modern world, whether capitalist or communist. The dominion of "objective reality" is a fundamental attribute of the civilizations of the West.[44] And that dominion is the dominion of the "I," the foundations of which the Western civilizations are consciously or unconsciously committed to consolidating, convinced (wrongly in my opinion) that the sense of the "I" and its continuity are the very basis of the personality and of a realistic adaptation to life.

The rational "I," product of a hypertrophied intellect, is a "defining-I," to borrow an expression from the contemporary German philosopher K. Dürckheim, an "I" that finds security (pseudo-security) only in strictly defined concepts, a "classifying I," an "I" held prisoner by space and time, an "I" avid for compartmentalization, separation, specialization, always in quest of "plans" set in advance, of "goals," of "objectives" to be attained, of "measurements," of "tests," of statistics, all excellent things when applied to the physical sciences and to the portion of human sciences dealing with social man, external man. But there is nothing more "unrealistic" and more dangerous for the human individual than overly strict definitions, classifications, specialization, compartmentalization, plans, goals, objectives, measurements, tests or statistics when one claims thereby to penetrate the domain of the deep realities of the human psyche.

The "death of the I" and the dialectical unfolding of a new kind of consciousness (the "re-birth" of the mythical religions) are the fundamental conditions for a return to the lost reality, a return to a view of the real and a mode of life that are realistic and balanced. From this point of view, the suggestive truths of the Koran—and those of the Old and New Testaments and of any other similar text—are not behind us; they are ahead, in a future that the human race is not yet sufficiently evolved to have reached at present, but towards which it is headed in a movement of slow dialectical evolution in which the rationalizing entropy itself, a quantitative entropy that is constantly worsening, calls out the qualitative and intuitive reaction—a psychobiological defense reaction on the part of the threatened race. That reaction announces and prepares for the blossoming of a new humanity, with transformed consciousness as a permanent "mutation," ready first to experience, then to conceive reality as one and all-encompassing, as neither "material" nor "spiritual" and meaningless except in the situation, when it expresses the experience of the human individual. A reality woven of matter and spirit, of science and the "imaginary," of reason and intuition, harmonized, reconciled at last in a race over whose head, as never before in its history, hovers the threat of destruction.[45]

"We have less than ten years left to save the world," Aurelio Peccei, the founder and president of the Club of Rome, wrote in 1979.

Notes

1. M. Eliade, *Naissances Mystiques* (Paris: Gallimard, 1959), 10, 25. The third sentence is emphasized in the text.
2. Ibid. Emphasis mine.
3. P. Diel, *La Divinite* (Paris: Payot, 1971), 60.
4. M. Eliade, *Images et Symboles* (Paris: Gallimard, 1952), 14.
5. J. Chavalier, *Dictionnaire des Symboles* (Paris: Laffont, 1969), Introduction, XXIX.
6. G. Gusdorf, *Mythe et Metaphysique* (Paris: Flammarion, 1953), 224.
7. Ibid., 228.
8. C. Jung, *La Guerison Psychologique* (Geneva: Librairie de l'Universite, 1979), Note 1, 27. Emphasis mine.
9. E. Cassirer, *La Philosophie des Formes Symboliques* translated from the German, (Paris: Minuit, 1972), Table 2, 31.
10. Gusdorf, 262.
11. Ibid., 260.
12. Superconsciousness: As it happens, I use this word in a very different sense from Diel's. Cf. J. Lerede, *Les Troupeaux de l'Aurore* (Myths, creative suggestion and awakening of the superconsciousness), Boucherville-Montreal: Editions de Montagne, 2nd ed., 1983); distributed in Europe by Dif-Edit, 96 Blvd du Montparnasse, Paris 75680, or Delachaux et Niestle, Lausanne, Paris.
13. Synergy: association of several organs or functions in the accomplishment of a task.

14. Cf. P. Diel, *Le Symbolisme dans la Mythologie Grecque* (Paris: Payot, 1970).
15. G. Durand, *L'Imagination Symbolique* (Paris: P.U.F., 1968), 109.
16. H. Corbin, *En Islam Iranien* (Paris: Gallimard, 1971–1972).
17. Eliade, 24, 17, 18. Emphasis in text.
18. C. Jung, *Introduction a l'Essence de la Mythologie*, translated from the German, (Paris: Payot) 215.
19. Gusdorf, 20.
20. H. Corbin, *L'Imagination Creatrice dans le Soufisme d'Ibn Arabi* (Paris: Maisonneuve, 1954), 13.
21. Durand, 30.
22. Contribution to the 13th International Conference on the Unity of the Sciences, Washington, 1984.
23. Durand, 6.
24. Durand, 23. Emphasis in text.
25. P. Ricoeur, *Finitude et Culpabilite* (Paris: Aubier-Montaigne, undated), 23.
26. Eliade, 23.
27. Van Der Leeuw, *L'Homme Primitif et la Religion* translated from the Dutch, (Paris: Alcan, 1940), 120.
28. For a detailed presentation (supported with specific arguments) of the hypothesis that the prehistoric caves were sites of initiation, see Lerède, 57.
29. Mythical thinking, *intuitive* thinking; there are many contemporary scholars of prehistory, mythology, or ethnology, such as Leroi-Gourhan, Leenhardt or Levi-Strauss, who have emphasized this *fundamental* trait of mythical thinking and mythical consciousness in general.
30. Gusdorf, 92.
31. A. Leroi-Gourhan, *Le Geste et la Parole* (Paris: Albin Michel, 1965), Table 1, 271.
32. Ibid., 293.
33. Ibid., 275, 287, 293.
34. 273, 275.
35. Ibid., 292.
36. Gusdorf, 92.
37. Ibid., 93.
38. Ibid., 93.
39. A. Veragnac, *Civilisation Traditionnelle et Genre de Vie* (Paris: Albin Michel, 1948).
40. A viewpoint I constantly put to the test in my psychotherapy practice.
41. Cf. Aurobindo, *Le Secret du Veda* (Paris: Cahiers du Sud, 1954).

42. "Except a man be born again, he cannot see the kingdom of God" John 3:3. See also the Epistle to the Romans 5:4, Epistle to the Colossians 2:12, etc.
43. K. Durckheim, *L'Homme et sa Double Origine* translated from the German, (Paris: Editions du Cerf, 1983). "When consciousness of the real is dominated by *reason*, everything that does not fall under its jurisdiction or surpasses it . . . becomes unreal or troubling and is laid to the account of simple faith, fantasy or metaphysical speculation."
44. The dominion of "objective reality" is a fundamental attribute of the civilizations of the West.
45. A reality woven of matter and spirit, of science and the "imaginary."

Discussion

Jean Lerède

Michel Maffesoli: The theme of initiation that you have just presented strikes me as important for what we are discussing. But I would personally like to make a small detour by way of myth. I feel, and here I refer to what Mademoiselle Cousineau said yesterday, that we cannot speak of an eternal myth. Myths, like values, are constantly changing. I know that the nineteenth century was rationalistic and yet had a whole series of myths, like the myth of the revolution. It seems to me that if we forget that there is always some modification, some transformation of myth, that we run the risk of falling into an attitude—and of this danger we must beware—which *stricto sensu* is reactionary, with no political connotations in this word, that is, reactionary in the sense that it only appreciates the past; or this other danger, shared by many contemporary intellectuals, of interpreting the myth as a catastrophe, or rather, in terms of a catastrophe. I wonder, and I am speaking here as a sociologist, if it would not be helpful to consider that, since myth is always modified, it may exist in another form today, that is, that we still have a certain number of rites and initiations even now. The development of a group structure in our present-day societies, which I personally call a new tribalism, refers us once again to the rites of initiation.

Thus I ask this question: when you speak of some balance which is now lost, aren't you forgetting that there are initiations and rites that are still operating? A whole series of research today is working on clarifying this point. These new rites, which refer to structures, that is, to small groups, spell the end of individualism, or, more precisely, the end of the principle Jung called individuation, which enlightened the whole Middle Ages and the Renaissance. The new structure would be more of a "group" order than of individuation.

I would like, quickly, to ask a second question, which concerns hyperrationalism. This term, remember, was used by Charles Fourier in the nineteenth century, but he saw it as the junction of reason on the one hand, and, on the other, its parameters, which pure rationalism itself had fled, like a dream, imagination, or even the idea of play. Today, however, and we can find flagrant examples of this, and for this reason our meeting on the imaginary is especially interesting and relevant, today we see a search for a new hyperrationalism, that is, something which can once more join together reason and its parameters, which until now have been forgotten or suppressed.

Jean Lerède: I am in complete agreement with what you have just said, and which nicely completes my talk, which was necessarily brief. You have given more specific details that I would have liked to give. I agree with you, and even though I say that initiation has been a lost phenomenon, we can now see that it is coming back, and really coming back in a great rush. It is reappearing in diverse forms, not only in the grand, traditional institutions, but more in the small groups you mention. It is reappearing, and from my personal experience I see it in seminars on creativity; and the members of these seminars see this phenomenon and identify it for what it is. These seminars themselves are real initiations, brought about by modern means.

Michel Maffesoli: Yes, because in big cities like the big American cities that in some respect make up the paradise of contemporary social development, these cities are no longer metropolises, but megalopolises, where we can observe the juxtaposition of groups; and there is a true phenomenon of initiation before getting into each of these groups, be they groups of consumption, pleasure, or ideology.

René Berger: As for me, I would like to make a few uncivilized remarks because I find them, at the moment, so distinguished. First of all, there is the idea of some lost reality, which assumes implicitly a "true" reality, which we would now be recuperating; and for which recuperation you give, if I can put it crudely, "recipes" . . .

Jean Lerède: I protest your terminology of "recipes."

René Berger: Please let me finish, you can contradict me later. Now! I will ask the following question: on the one hand, there are those who feel a great nostalgia for this paradise lost, as we all do. But on the other hand, I look around me and ask: how is present-day society operating? I return to the definitions that were suggested of megalopolis or metropolis, as you wish. I note that rites of initiation still exist. They exist, but in another form, and this we seem to have misunderstood. What is the only rite of initiation we all undergo? It is getting our driver's license [laughter in the room]. You may laugh, but it is a fact that with my driver's license I am initiated into the world of traffic, which is one of the major behavioral patterns of our society.

Second, the idea of myth. No doubt, it is a wonderful thing to speak of myth and to refer to Mircea Eliade, etc.; but there is a powerful myth today, and that is the myth of advertising. We can say that it is a mutilated or impoverished myth, but it is the only myth that operates in daily practice. You are French or Canadian: take the Renault automobile company, which yesterday decided to spend six million dollars to advertise their new car on American television—another new car! Can you tell me what form of art could benefit from a government, or an enterprise, from such a budget for advertising? Thus I say that myth exists in our time, and that myth is advertising.

Third, in these rites of initiation you spoke of, and you might think this is funny, too, I think these rites go on today. I spoke about your driver's license, and now I would like to mention shopping. Shopping is today, I believe, a rite of initiation. Whenever I can go to the supermarket and take out my credit card, since spending dollars in the United States is looked down on, I am in fact one who has been initiated. With my credit card I have become part of a certain world, of a symbolic world. It is, if you will, a parasymbol, or a degraded symbol—it doesn't matter which—but it is certainly a powerful symbol of ritual.

Fourth, death and rebirth, this thing you find everywhere . . .

Jean Lerède: It is not a "thing"!

René Berger: No, but I am saying that fast, because I must take shortcuts because of time, so I caricature those shortcuts. Death and rebirth is today either having money or not having money. Dying is not having any money, and having money is being reborn.

Fifth, hyperrationality, this deplorable characteristic of our time, as you state. We can say in any case that it works, and for me this hyperrationality has been translated into the power of technology. Hyperrationality is the existence of machines, which determine reality, which make things work or not work. As for this world, then which I can personally accept completely, I think that no one has the right (and I am not accusing you personally) to say, in the name of a certain nostalgia, that this world should lead us to a nostalgia for the past. On the contrary, we should confront our world as it is.

My conclusion is an answer to this question: should we favor a return to this inner poetry, to this paradise lost? I appreciate your talk very much, and this is my personal opinion: I think we should confront the realities of this world, a reality we can call artificial or superficial, or shallow, but in any case, the reality in which I am living, or we all are. This reality merits our consideration, if not more. Forgive me for having seemed brutal. I am really tenderhearted, but I don't know myself well.

Jean Charon: I would like to take advantage of my position as "chairman" to answer René Berger, even before letting Jean Lerède respond. I understand your point of view and for the most part I agree. But the initiation you are describing is not, I think, and Jean Lerède can confirm this in a minute, the initiation that Jean Lerède was talking about. This initiation is a quest for a dimension that does not link us directly with the reality I would call "outer" reality, or our daily reality of the world we live in. I could also call it spatial-temporal reality. Automobiles, advertising, drivers' licenses, all those things are part of an external, daily reality. Initiation tries to help us reconnect with a dimension I would call transcendent. We can ask what transcendent means. It is that dimension that links us to our own inner reality, or in short, our spiritual reality. . . .

René Berger: Jean Charon, I ask your friendship in believing that I understood perfectly what you just said. I also understood that Jean Lerède was speaking to this inner dimension you are talking about. But I interpret these things differently, and I give an initiatory priority to our adapting ourselves to our contemporary society. You may think I am confusing things here, but for my part I believe there are confusions that must be made, by necessity, today.

Jean Lerède: Overall, René Berger, I fully agree with what you said, but, as Jean Charon noted, I was really addressing another type of initiation and other myths.

8. The Paradox of Values

Diane Cousineau

Our entire notion of objective reality, of a reality independent of our subjective perception and possessing an existence of its own, now appears to be undergoing radical questioning.

The East has always had an extremely paradoxical notion of reality. "The Eastern mind," as Professor Izutsu will remind us in his contribution to the Conference, "begins philosophizing from the starting point of a radical and thoroughgoing negation of the ontological reality of the empirical world." In Eastern philosophy, in effect, the things we perceive as existing in our empirical world are and are not at the same time; they are what they are at the same time they are everything they are not. In other words, nothing has an essential identity; everything is simultaneously itself and its opposite. Now that Eastern thought is penetrating further and further into the West, it is reasonable to suppose that this phenomenon has had some impact on the currently changing mentality.

Moreover, an entire current in contemporary Western science, involving in particular the latest developments in physics, has played its part in profoundly shaking our conceptions regarding the objective nature of material reality. For example, the "bootstrap" theory of Geoffrey Chew (University of California at Berkeley) notes so intimate an interrelation between the elements of the real that he goes so far as to say that each particle is less important for what it is in itself than for the relation it bears to all the other particles. In other words, nothing can be considered in isolation as a defined entity without at the same time taking into consideration all that it is connected with, i.e., all that is not that defined entity.

John A.Wheeler, director of the Theoretical Physics Laboratory of the University of Austin, Texas, tells us in his "participation-observation" theory that there is so close an interaction between the reality observed and the subjectivity of the observer that one cannot describe the one while neglecting the other. The so-called "objective" reality and the subjectivity are inseparable, as if they were after all two faces of one and the same thing.

Finally, Jean Charon, with his theory of "complex relativity," goes even further in that he discovers an imaginary space/time associated with all material reality and makes a model of these two faces of the real, objectivity and subjectivity, reality and the imaginary.

These various scientific theories, particularly those of Wheeler and Charon, do much more than represent subjectivity, or the imaginary, as a dimension parallel to concrete reality. They displace, so to speak, the center of gravity of reality itself, so that the heart of reality is now located at a sort of meeting point between the world of the imaginary and the world of observable forms. The traditional dichotomy between objectivity and subjectivity, or, as Parmenides puts it, between the "Way of Truth" and the "Way of Opinion," now seems to seek resolution by recognizing their necessary interaction, which creates reality as such. Therefore, the imaginary in such theories can no longer be understood as a fiction that contrasts with concrete reality, or even simply as the eye that looks at reality and gives us a more or less distorted view of it, but as an actually intrinsic part of reality itself, essential not only to its representation, but to its composition. The imaginary can no longer be discredited as being "that which is not anchored in reality," since, on the contrary, reality itself turns out to be deeply anchored in the imaginary.

On the other hand, in the field of psychology, the discovery of the unconscious and the continual deepening of knowledge about its nature and functioning alerts us to the extent each of our conscious actions stands in a dynamic relationship to an entire unconscious content and gives rise to and simultaneously contradicts those actions. The unconscious is also the seat of the imaginary in the "noble" sense of the term, in other words, its creative dimension as opposed to the pejorative and reductionist concept found in most theories of the imaginary in the West.

Yet, it is understandable why the term "the unconscious" seems to some inadequate. While it covers an extremely live, rich, and complex area of psychic activity, it remains semantically tied to a negative concept ("absence of consciousness"). Some authors, even while continuing to use the term "the unconscious," feel the need to state reservations for its value and to warn against possible pitfalls in using it. "To use this word to designate the domain of the imaginal, behavior that is morally inferior or below any cultural norms, instinctive reactions of liberation, a causal agent that 'sends' dreams and to which one can turn to get an opinion is to break down categories," James Hillman tells us. For this reason, he suggests that we should not only be wary of the term, but should reexamine our notion of what we call "consciousness." This reexamination, contrary to the positivist conception limiting it to the psychic phenomena called "conscious," should encompass a large part of what we have up to now relegated to the domain of the "unconscious."

Be that as it may, it is more and more clear that psychic reality, like reality itself, can no longer be limited to the domain of the "conscious," which represents only a portion of it; moreover, even that part of psychic reality that manifests itself on the conscious level turns out to be closely linked to its "unconscious" counterpart. Just as in physics, observable reality is revealed as laden with extremely complex, invisible ramifications, ramifications that are not only an integral part of it, but determine it and give it existence. And in light of the positivist, objectifying mentality that has long prevailed in scientific and psychological research, it can be said that these new advances represent something quite other than a simple step forward in our way of perceiving and understanding reality; they constitute a radical inversion of our very notion of reality and our comprehension of the world and of what we are as psychic entities. We are experiencing a revolution, in fact, much more than an evolution.

Is it surprising that such a phenomenon coincides chronologically with what we generally recognize as a "crisis in values" now shaking our Western world? Can we, after all, call into question our ideas of objective reality, whether material or psychological, without at the same time, whether we want to or not, questioning the objective reality of what we call our "values"? Can such a theoretical examination not have repercussions on the concrete, practical level; can it not finally affect us in the daily choices of our lives and in our mentality? If there is, in fact, a relation of cause and effect, one could probably just as readily reverse the terms; for while it is true that developments in scientific knowledge are bringing about changes in the mentality of the times, it is no less true that they are themselves brought about by changes in mentality. Scientific discoveries are both the motive force behind and the evidence of the psychological evolution of a society. But whatever the direction of this cause and effect relationship, if we define values from a psychological standpoint as being the realities consciously preferred by an individual or a society, it would seem inevitable that our notion of value should be subject to the profound questioning that is now affecting our entire notion of reality. Hence, what we call the "crisis of values" may in fact be interpreted as one of the many manifestations of a much wider crisis that I take the liberty of calling the "crisis of reality."

As in the case of material reality, the question thus arises whether "values" are objective entities, whether they (or at least some of them) exist by themselves in the absolute, independent of our choice, adherence, belief, or our conscious or unconscious decisions. A portion of this debate, which ultimately touches on the ideas of "truth," "good," and "evil," belongs to philosophy and even spirituality. What I will point out here are certain avenues for reflection that depth psychology opens up for us, basing my discussion in particular on Jung's theory of the archetypes and more generally on the mythical imaginary, which one can view, moreover, as an expression, a manifestation of those potential forms that constitute the archetypes.

Jung tells us that all psychic reality, conscious or unconscious (and hence each of our values), has its source in archetypes. Now, the nature of an archetype is to represent a unified but nonetheless complex psychic whole made up of components that may be paradoxical and even contradictory, but are all intrinsically linked one with another, as by a bond of interdependence. None of the components of an archetype can be evoked without automatically evoking everything in the archetype that is other than the component, in other words, its archetypal complement, including the part of its complement which is diametrically opposed to it.

The archetypes, being as such "potential" forms or "empty" forms, by definition can only be apprehended indirectly, for example, through the avenues of myths, or of dreams, which give them existence while manifesting their content. Myths appear, so to speak, as the expression of the potential relationships of the contents of the archetypes. Since myths are imaginary forms that are close to their archetypal sources, they serve, in part, as illustrations of the inherently paradoxical nature of archetypes. A myth tells a story about a figure who incarnates one component of the archetype. But the mythical figure and the tale form an inseparable whole. Just as Hercules is inconceivable divorced from his labors, no mythical figure, taken in isolation, has meaning. He acquires it in relation to and even in confrontation with other mythical figures or elements. Considered in isolation from the whole, he is an abstract entity. He is revealed and becomes a true active force due to his dynamic contact with his mythical world, where he finds allies, opponents, even enemies, all essential to the development of his identity. Thus, one can say of a mythical figure that he is what he is as much by his relationship to the elements of the myth as by what he is in himself. Furthermore, the mythical figure in himself is generally ambivalent; more often than not he presents a double face, positive and negative. By way of example, let us take the goddess Athena, builder of cities, goddess of civilization, the bearer of peace depicted with an olive branch in her hand. But this is also the goddess who sprang fully armed from the brain of her father, and whose weapons, moreover, are not purely defensive. Athena is also she in whose name wars are declared, who thus brings about the destruction of cities and ultimately threatens civilization. Athena is what she is and at the same time is her own opposite.

If we consider the psychic totality of an individual, bearing in mind his unconscious as well as his conscious dimension, we realize that the human psyche functions precisely according to the dynamic and paradoxical model that characterizes the mythical world. This should not surprise us, since the mythical imaginary is not only a product of the psyche, but its most "primitive" product, its first formulation. In myth, the human psyche is telling about itself; it is therefore natural that one mirrors the other. Whatever value an individual chooses to prefer at the conscious level will be in his unconscious a swarm of impulses that support, upset, or even squarely contradict the conscious statement. We may then compare a value to a mythical figure and note that in isolation from the psychic whole it is, like the mythical figure, an abstract entity, and becomes an active force not in spite of the encounter with that in the psyche which responds to it, questions it, resists it, but due to that encounter. It is precisely all this psychic material that lends existence to value, that undergirds it and gives it energy. Hence, a holistic view of the psyche that comprises conscious and unconscious shows us that values are not absolute realities, but entities eminently relational and deriving their existence precisely from their paradoxical nature.

However, rational consciousness, "objectifying consciousness" to use the term of the German psychotherapist Karlfried Durckheim, has trouble accommodating itself to so paradoxical a reality. Its function being to define things, hence, to isolate and delimit them, it can only connect things that are rationally compatible. To the objectifying consciousness, paradoxical relationships appear problematic and antithetical relationships appear threatening or quite simply impossible, since they are the very negation of what has been stated. This is why we often tend to want to master the paradox by eluding it; what seems by nature to contradict the consciously preferred choice is devalued as belonging to the domain of "evil" or "anti-value," or denied, or carefully buried in the unconscious. We thus try to control human consciousness by imposing on it a forced correspondence with a single one of its dimensions, to the exclusion of the others.

The problem is all the more acute in that values, to be incarnated collectively, demand recourse to what we might call "symbolic models." A religious ritual, for example, is a symbolic model that incarnates an aspect of the spiritual life of a group. Marriage is a symbolic model of love between men and women while a particular political system represents a symbolic model of the life of a society. Symbolic models are in themselves representative forms of values, somewhat the way we say of scientific theories that they are representations of the real; but since they are simpler to discern, define, and hence control than the values from which they emerge (but which always transcend them), the objectifying mentality readily reaches a point where it identifies the value with the preferred symbolic model, thus confusing the real with the representation. One can even say that in such a situation the model loses its symbolic quality and becomes an objective model. When conflict arises between the dynamic of a value and a too rigid model, it is not rare for an individual to show loyalty to the model, the representative, exterior form, rather than to the value, the inner impulse the model symbolizes. The model, of course, has the advantage of being better defined, more perceptible, and reassuring to the conscious mind by supplying it with an objective point of reference that leads to a definite position. But a split then arises between the model and the value, between the objective reality and its imaginary matrix. Maintaining such a split is like cutting a flower from its root, separating it from its nourishing plot of earth, and expecting it to bloom.

Since, moreover, the symbolic model generally represents only a specific aspect or a particular sociocultural conception of a value, it is almost inevitably jostled by everything that represents another aspect or conception of the value. It is even more likely to be jostled by all the other values that demand their place in our psychic landscape, demand to be accepted in turn, to be linked one to another. Whether we are talking of peace, liberty, growth, justice, love, or the value representing life itself, none of these entities can be separated from the others, for they are, so to speak, contained in one another and give meaning to one another reciprocally, even when it happens that they appear to conflict. They interpenetrate in the same way that various mythical figures appear in a tale in which only one of them is the hero. The mythical world, like the world of values, is made up of constant interaction between figures who take on new roles in new situations, from one tale to the next.

If, however, we can understand how a value connects with certain paradoxical realities, how do matters stand regarding that in the psyche which represents the opposite of the value itself? If I manage, for example, dynamically to reconcile love (which may be defined as the force of attraction, the desire of two beings to unite) with my desire for freedom, inner growth, and personal identity, does it follow that I will be able to reconcile that force with its contrary (which would then be a force of resistance or even of repulsion)? At first sight these two impulses seem to negate one another. It is precisely the simultaneous existence of these two contradictory impulses, however, that creates a dynamic space into which the paradox can be fit; it is this contradiction that, once accepted, allows me to love and at the same time grow, be free, be myself, and reinforce thereby the quality and vitality of my love. The presence of the opposite is not merely necessary for the vitality of a value; it is actually the *sine qua non* of the value's very existence. Love, as a psychological reality, demands for its existence the simultaneous presence in the psyche of attraction and repulsion. Without the presence of one or the other of these two contradictory components, one would end up with the annihilation of love.

Every value is similarly in a necessary relation with its complement, whence it derives its vitality, its character, its special color, and with its opposite, which is the very condition of its existence. From this point of view, paradoxes, even contradictions, no longer appear as the expression of a deplorable condition in the human psyche to which one must adjust as well as one can, but rather as a fact inherent in the psyche and as the actual foundation of the dynamic of each of its components. Paradoxes are facts, therefore, that the individual is called upon not to conquer but to accept as necessary and legitimate. This acceptance, this *conjunctio oppositorum*, even seemed to Jung to be the essential condition for the development of the individual and the realization of his full maturity, the "self," i.e., the integration of the whole personality.

But by opening the door to impulses that contradict our moral choices, by considering them as legitimate and even necessary, do we say that everything in human consciousness is of equal value? Is welcoming the paradox, reconciling the opposites, to embrace nihilism on the moral plane? Such an attitude, on the psychological plane, does not deny the existence of values any more than new theories in physics deny the existence of material reality. Acknowledging the fundamental relationship between things and the ambivalence of each of them by no means signifies a flight from moral order into chaos, or a denial of concrete reality in favor of its imaginary dimension, but rather the recognition of the essential link between these two polarities that make up the real. Nor does it mean that values as such are arbitrary and that the decision to favor this or that component of the psyche does not matter. The great values such as love, justice, and respect for life, for example, appear to be the irreducible givens of the psyche by which life manifests its call to life. As such they could be called absolutes.

However, the meaning generally assigned to the notion of "absolute values" reveals in most cases a profound misunderstanding of the basic nature of values, which are the psychic forms emanating from creative consciousness. The human psyche, as it happens, is at one and the same time immersed in the world of primordial images and projected into the world of determinate forms; it is rooted in the imaginary and deployed in reality; it participates intimately in chaos, its matrix of creation, and in order, its creation. What we call "consciousness" is precisely the power that transforms the contradictory state inherent in the psyche into a creative relationship, a kind of thread stretched from chaos, which it is constantly trying to order, and then dismember so as to reformulate it, with the result that order and disorder reciprocally give meaning to each other. Consciousness, therefore, is not a state, but an act that springs from the encounter of these two psychic states in perpetual interaction, and values, as living manifestations of that consciousness, participate in this paradoxical dynamic.

When, however, we speak of absolute values, we often tend to confuse the value itself, which is a psychic form in constant gestation, with a symbolic model, which in itself is nothing but a representation, a momentary product of creative consciousness (even when it succeeds in lasting for centuries). To adhere to symbolic models as absolutes is to confuse the real with its representation, the thing with the conception one has of it; it is to betray the creative process itself in favor of the created. Let us listen again to Karl Dürckheim. "Whatever is alive lives only by the process of becoming. But what has already come to be is always both a pre-condition and an obstacle for what is yet to come. That which is solidly established hinders life in its constant mobility, and the fate of man calls him to engender and live this inner contradiction." If there exists an absolute in the world of values, it cannot be found in any specific form, but rather in that aspiration of consciousness to which values bear witness, an aspiration of chaos towards order, which gives it meaning, and of order towards chaos, which keeps it from petrifying.

This creative consciousness, which allows the psyche to be immersed in chaos without foundering in it and to project itself into order without congealing in it, is incarnated in the life of each of us due to the awakening (or rather reawakening) of the intuitive faculty and hence the accession to a "dual-level" consciousness, as Jean Lerède writes. Profound intuition guides the psyche through its own labyrinth, like Ariane's thread; it enables a being to accept, creatively, everything he bears in him.

Of course, it may be, and it is more than likely, that such an individual will depart from a certain image of perfection, if one understands by perfection correspondence to an ideal model, ordered in a precise and predictable manner. That is why Jung insists that "wholeness is not perfection, it is the integration of the being"; and he adds, "This is the way man becomes the difficult problem that, precisely, he is." A problem, yes, like any authentic creation. Indeed, the way the individual, thanks to his intuition, brings together the various components of his psyche, evades any attempt at advance control. No preestablished model can provide a solution, for the solution, if it is creative, will necessarily be unprecedented; by that very fact, no solution can be imposed on another individual. It may serve as inspiration, but never as an absolute rule. Being unprecedented, every solution that emanates from intuition, before it takes shape in consciousness, therefore remains by definition unpredictable, not only to others but to the individual himself; for it is the nature of the intuitive faculty to weave the web of destiny little by little, to generate in the conscious mind the general sense of a direction while unveiling the concrete stages only step by step. A being listening to his intuition comes face to face with himself, as a novelist comes face to face with the character he has created, a character who assumes life and substance only to the extent that he exemplifies the volition and conscious plans of his author.

If we look at what is now taking place in the world of values, we find that the models, not the values, are in crisis. What motivates the present generation is, on the deepest level, what has motivated preceding generations. Peace, freedom, justice, love, even love in the aspect of the couple, still (one could even say more than ever) represent the fundamental aspirations of individuals. What is changing is the way all these things are incarnated. It is as if values are increasingly demanding to be recognized in their complexity, in their dynamic correlation with the psychic whole, to be accepted by the conscious mind together with all that gives them life. The present generation is claiming the right to its paradox, even to its contradictions, and is thereby demanding that reality be re-rooted in the imaginary that supports and impregnates it. It is a claim for wholeness that is expressed in an anguished and often awkward way, to be sure, but that is how everything in the process of development usually expresses itself. At the bottom of it all there is, as it were, a mythical figure who is demanding, today and in each one of us, to reintegrate his mythical world, his story, his adventure.

On this integration of human consciousness depends not only the future of the individual, but our collective future, as well, Jung claims. "If an inner fact is not made conscious, it presents itself from the outside, as fate. In other words, if the individual remains monolithic and does not become conscious of his internal opposition, it is likely that the world will have to act out the conflict and be split in two." Indeed, any impulse, even any image, is much more to be feared when it is repressed than when it is acknowledged and accepted by the conscious mind. It is no less active for being repressed. Somewhere in the life of the individual it will come to claim its due and more often than not will do so with a devastating power commensurate with the force of the denial ranged against it. Do we not daily have before our eyes, moreover, a striking example of the danger we run from the state of psychic dissociation in which we are still for the most part trapped?

No civilization has been so dominated by the image of life as ours, to the point of wishing to efface from our consciousness everything that reminds us of death. Before the advent of our twentieth-century technological civilization, life and death were constantly and concretely acted out side by side; most children died at a young age, parents died in the midst of their families, and epidemics and wars regularly carried off members of the family. The image of death was present everywhere, an image that has become unbearable to our contemporaries. Now we suffer and die in hospitals, far from the sight of the healthy; our old people are hidden in nursing homes so that they will remind us as little as possible of what we want to forget. By all sorts of artifices we carefully repress in the depths of our unconscious that image of death that appears to us an affront to life. Yet it is this same twentieth century, clutching its cult of youth and denial of death, that has invented weapons capable of wiping out all life on the planet in a single instant. Consciousness of death would no doubt have made us more aware of the preciousness of life. Death, repressed into our unconscious, threatens us daily and globally; it has become omnipresent. The threat it presents to our civilization is the measure of the denial we have raised against it. And it is to be feared that no treaty, no alliance, will protect us if we have not made peace with that image from which we have cut ourselves off, but which nevertheless dwells in us. The future of reality itself depends on its being rooted in the imaginary, with all what that implies.

Discussion

Diane Cousineau

A member of the audience: If I understood correctly, you were trying to defend the point of view that both values and anti-values have their own reality. Since we cannot choose both of them at the same time, doesn't that mean that we are condemned to a sort of "mediocrity," because we find ourselves not being able to choose only black or only white, but, since we recognize the reality of both black and white, we are almost obliged to put ourselves between the two, that is, to resign ourselves to mediocre values?

Diane Cousineau: No, that is not what I meant. I intended to express the following notion, which has been part of oriental philosophy for thousands of years: there is a kind of ambivalence of values, that is, that a value, to be what it should be, must necessarily be supported and accompanied by a cluster of other values which uphold it, which sometimes contradict it, since in this cluster there is also anti-value, that is, contrary value. The ambivalence of value is found, as I indicated, on the level of archetype, that is, the deepest level of the psyche. When you want to live a value, you live it through a symbolic model, that is, a *lived* model; in that case, you can choose between black and white. But the point I wanted to emphasize is that in choosing such a symbolic model, you must not forget where its roots are. Its roots are in the choices that this model brings with it, but also in the ambivalence of the archetype that this model seeks to express; and this ambivalence allows us to live our own model, while being aware of what makes it itself, that is, the relationship between this model and all the other models and especially all the other values. I think this can be illustrated as physics has done (in the "bootstrap" theory) in considering particles of matter. A particle is what it is not only by its own structure but also by the relationships it has with all the other particles in the universe. You see, as long as physicists *forgot* the existence of these relationships with others they had a much more reductionistic idea of what a particle was. I showed, likewise, that values lived through symbolic models must be lived with the awareness that they are what they are and are worth what they are worth as much due to what they really are—they are white, for example—but also that they are white precisely because there are many other colors, a whole rainbow of colors, which give real meaning to the white model. It is, almost, the beginning of a tolerance of values that cannot exist in isolation, that is, that do not have an absolute existence but that owe their existence to all the relationships they have with all the other values.

Henryk Skolimowski: Thank you for this wonderful talk, beautiful and generous at the same time, but it seems to me so generous that it really could not be true (laughter in the room). Since you encourage us to recognize a paradox of values, I would like to compare it to the paradoxes I proposed in my own talk.

Diane Cousineau: Then I will try to connect your paradoxes with mine!

Henryk Skolimowski: First, an objection, then, a question. You say that our time is not so much undergoing a change of values as an opening of the framework in which we live those values, and that such an opening is being demanded by our contemporaries. You emphasized that great aspirations, such as justice, freedom, and love, are as much sought after today as before. But are those ideals really the same as values? If you look at all the evil modern technology imposes on us, and the values associated with gross commercialism, then we realize that many people would like to *change* these values.

Diane Cousineau: Yes, but I think that the values themselves "inhabit" these aspirations, and not the way in which these aspirations are lived. I think what happens in our modern world is that we want to make these aspirations real, but we suddenly see how far we are from what should be the realization of these aspirations. But the basis of our aspiration is always there, and we always aspire to values that are the fundamental values of the past.

Henryk Skolimowski: Would not this be a kind of search for a paradise lost, and a search we will always have since we are always unfulfilled?

Diane Cousineau: I do not think this idea of paradise lost is typical of the twentieth century. But you can always speak of this feeling of paradise lost because the values I referred to are, in fact, transcendent; they are founded on an archetype and must always be questioned. It is a quest, as I said, which starts from what has been ordered, to create a new chaos, then to discover a new order afterwards. This is the great impulse at the heart of all value. It is transformation. What we do to remain true to a value will never be the same as the value itself.

Henryk Skolimowski: Let us try to get to the bottom of the problem a little more, and here is where I come to my question. At the beginning of your talk, you spoke of values associated with the reductionist or mechanistic view of Newton's time. At a certain point, because you spoke of relativity and relative ideas in physics, I thought you were going to arrive at the relativity of values, which would contradict the absolute values of the last century. But I seem to note that you are going back to a desire for absolute, not relative, values. Your own point of view, I feel, is not clearly expressed: do you lean toward relative values, or do you think that the notion of absolute value can be logically defended, considering the state of modern knowledge, especially of physics? You also stated, and I see it in your paper, that if we look for something connected to absolute values, then we must look for it in our aspirations, and at this point in your talk you seem to change the context surrounding the notion of value, which you now assimilate with the notion of aspiration. This bothers my understanding of what you want to show us. If you believe that love is an aspiration, then I agree with you; but then, is it a value?

Diane Cousineau: What is relative is the way we live values. If you look at what could be called the tree of evolution, you can see that certain major aspirations appear repeatedly, but the way we live these aspirations, especially today, changes and varies. One person is wrong if he wants to tell another how to live his own aspiration. He is wrong because he is confusing what is changing, which is the lifestyle, with what is rigid and invariable, which is the aspiration itself. The way we live a value is relative, that is, the way we embody this value.

Jean Charon: I think that, as Diane Cousineau has mentioned, we can differentiate between the value and the embodiment of this value in what may be called a symbolic model. For example, love can be lived out in one of these possible symbolic models, which may be religion; but this symbolic model tends to think itself absolute so that, generally, it develops only *one* aspect of this value, by isolating this aspect from all the allied or even contradictory aspects of this value, from the other aspects that underlie it and give it its own existence. In such a way, the symbolic model, if it is lived from this narrow point of view, loses all the rich fullness of this value. It is a little as if, and here I'll take an image used by Diane herself, you wanted to make a flower live by cutting it off from its roots. Of course you know what will happen in a little while. The type of existence you gave to this flower has, in fact, sentenced it to death. In the same way, we must, I think, be cautious in constructing symbolic models; we must be aware of the whole range of values which underlie the symbolic model we want to embody.

Françoise Bonardel: I was moved by your paper, and especially by the idea of lived reality, which underlies your thoughts. I would simply like to ask one question: if everyone, or almost everyone, today agrees on the importance of the paradox for creativity, do you think that this creative power, this paradoxical dynamism, is only linked to the difference in potential of two antagonistic realities, as Jung suggests, or is there some undiscovered polarity which should intervene to animate and complete its development? I feel there is something a bit ambiguous there. How can we get out of this paradox if we really don't know the resolution of the paradox? At the same time, we really don't know this resolution, or else we would have resolved the paradox already.

Diane Cousineau: I appreciate your remarks because it seems that if there really appear to be opposites in the beginning, then admitting that those opposites could be resolved would create another entity, which, by its very existence, would create another contradictory entity.

Françoise Bonardel: At that point, wouldn't it be the role of the archetype, which is not only a form but also a polarity, to stimulate them to go beyond this contradiction; this is an entity, remember, that Jung compared to the alchemist's stone. Is there not a "mystery" there in wanting to go beyond opposition? And if this contradiction were reduced, would there not be a risk of making this resolution of paradox a simple paralogical formula which would take away its mysterious quality which is, perhaps, inherent in all formulas?

Diane Cousineau: Yes, and I think that this reconciliation of opposites is mysterious by nature. A true reconciliation can only be accomplished, I feel, through an experience of *initiation*, which would be on a whole other level of consciousness. We need a radical change of consciousness, that the mystics call an interior death, that Orientals call *sûnyatâ*. And on that subject I refer you to the wonderful paper by Isutzu at our present meeting. This reconciliation can be accomplished only by such a change of consciousness. It would be dangerous to try to rediscover these polarities, these different elements of consciousness, with our observing mind, because our old rationalizing, objectifying mentality is by nature fragmented. In this case we discover only a fragmented world and as such we ourselves remain fragmented beings. Certainly, the sense of the mysterious does not give us an answer to mystery, but does allow us to proceed in the right direction, that is, never to think we have found something, but to transform ourselves unceasingly, not to structure ourselves into immobility. Because immobility is death, is it not?

9.
The Ontological Ambivalence of "Things" in Oriental Philosophy

Toshihiko Izutsu

The expression "ontological ambivalence" in this paper refers to one of the fundamental problems of Eastern philosophy, the ontological condition of things amidst the empirical world, and particularly the attempt to establish if they are "real" or "not real." This central problem characterizes most of the great philosophical traditions of the East in a very remarkable way.

First, the idea with which we are dealing can be succinctly presented in the form of a basic proposition, each thing *is* and *is not* simultaneously. Thus formulated, the idea seems to reveal a highly paradoxical nature. In fact, the postulate means that all that is perceived as existing does not exist in reality. It also means that all that is perceived as such or such a thing in the empirical world, actually is not such and such. *A* is *A*, but is not *A*. *A*, while being indeed *A*, is not *A*. *A* is "nothing," and yet is *A*. What is the foundation of such an apparently illogical statement, which looks like a shocking contradiction in terms, and why is it offered? Let us try to elucidate this problem theoretically, from the point of view of Eastern philosophy.

The Eastern mind, we could say, begins to philosophize, when it does philosophize, from a radical and resolute negation of the ontological reality of the empirical world. There are, of course, considerable exceptions, such as the primitive Confucianism of China and the Nyāya-Vaisesika of India, but the majority of the main traditions of Eastern philosophy are characterized by a special type of ontological negativism, which consists in annihilating things, or we could say, in "de-reifying" the *reses* with the corollary of Mahayana Buddhism's famous statement: *sûnyam sarvam*, "all things are voids."

This concise but strong statement summarizes the entire philosophy of Mahayana Buddhism, such as it is represented by the *Prajnāpâranitâ* Sutras. The question however is knowing what that means. At the level of popular thinking this postulate was given, in a very comprehensible way, nihilistic interpretation. By way of example, here is a passage from a Buddhist Scripture:

> All things come to be and become nothingness. Nothing lasts eternally. (Things) are as phantoms, clouds and flashes of light. None of the things continue to exist. To tell the truth, none of them last even an instant. All that exists is actually non-existent. It is only a dream, a puff of warmth in the air.

Similarly, the master of Zen, Sêng Ts'an (?–606), the third Patriarch of Zen Buddhism in China, said:

> *Dreams, illusions, flowers fluttering in the air,*
> *Why are you so impatient to catch them?*
>
> (*Poems on the Mind*)

By categorically negating the reality of all things and by calling them phantasmagoria, Eastern philosophy attacks the world conception based on the common sense of the common herd, which can be described, according to philosophic terminology, as a world conception resting on the foundation of a naïve realism. The naïve realism, which Eastern philosophy combats, must be understood, in the present context, as a popular ontological belief in the objective existence of the outside world, the physical world being conceived as a solid whole consisting of innumerable "things." In this conception, the empirical world thus becomes an ontological space filled with "things," each one of them being a physically *real* substance manifesting itself by *real* activities.

It is characteristic of these substances to be "essentialized"; that is, each one of them is endowed with an ontological core, an "essence," around which it crystallizes and by which it distinguishes itself from all others. Each thing, in this way, is irrevocably itself. And that is where we find the secret of the supremacy of the identity and non-contradiction laws, which rule over the so-called logical thinking of the common man, who himself chooses to remain attached to the common-sense philosophy of naïve realism. It is a world conceived according to the formula: *A* is *A*, *A* can never be non-*A*. In other words, all things are—each one in turn—closely protected by their own "essential" limits against all other infiltration of their domain. Eastern philosophy combats such a conception of the empirical world by trying to "liquefy," so to speak, the "essential" solidity of all things.

I quoted earlier some passages from Buddhist texts, showing how Mahayana Buddhism in general thinks of the empirical world as a dream. Classical Taoism, in this respect, seems to agree completely with Buddhism. There are, in fact, several famous passages in the *Book of Chuang-tzu*, where the Taoist philosopher seriously tries to convince us that the entire world of existence is nothing but a "great dream." The Vedânta seems to express exactly the same idea, through the key expression Mâyâ, which means that all the things in the world are only just as many phantasms-*Mâyâ*; whereas the Irfanic philosophy of Islam seems to talk about the totality of the things of the empirical world as being an apparition of the *Khayâl*, that is of the cosmic imagination.

Therefore, we register everywhere in Eastern philosophy basically the same idea-image playing a decisive part as the first starting point of several forms of negative ontology, negative in the sense that it begins by negating the reality of the empirical world. As we have seen, all things, even the "strongest" material things, are treated in this kind of ontology like dreams, illusions, or errors. All things, thus, are "absences" and "voids."

The idea according to which the entire world is only cosmic phantasm-agoria, an ontological illusion on a large scale, can leave man in a boggling pessimism as witnessed by the development of popular Buddhism. However, the real philosophical implication of the idea is something quite different. By trying to clarify this point in the course of the analysis that follows, I hope to bring to light the true nature of Eastern philosophy's notorious negativism.

First of all we should observe that the most outstanding trait of the human mind (or of human consciousness) is that it is, by principle, polarized between the realm of the subject and that of the object. The subject-object polarization of consciousness, at least as it appears on the surface, induces consciousness to identify entirely with its own subjective realm by excluding what is objective, and to establish itself as *ego*, existential center of all the noetic (intellectual-abstract) experiences. The objective realm of consciousness is, therefore, so to speak, removed and transformed into an "external" reality, existing by itself and in opposition to the "ego."

In such a situation, the mind can only act as an objectifying consciousness. Inevitably, all it does is objectify everything it finds in the "external" world. This makes up the basic structure of the *noesis* (intellectual abstraction), as it is commonly understood. This is also the meaning of the assertion that consciousness is invariably and necessarily "consciousness-of" and never pure and simple consciousness.

"To objectify" something is to "essentialize" it, for we cannot, in the sphere of intellectual abstraction, grasp anything at all as an "object" unless we recognize in this thing an "essence." *X* is recognized by cognition to be an object, that is to say as such and such a thing (a flower, for example) uniquely because the mind rediscovers in *X* certain essential characteristics (for example, the fact of being a flower) that permit *X* to be a member of the group of things designated by the name "flower." The mind can (and ordinarily must) function as "consciousness-of," because it everywhere perceives "essences." "Consciousness-of" simply means consciousness of an established object via its essential determination. Consciousness devoid of an object becomes in this context a meaningless expression.

It is a similar "objective" appearance of the world (which is in fact a "subjective" product of the "consciousness-of") that the Eastern negativism tries to sweep away with one motion by stating that all phenomena—be they things or facts (events)—are only dream-images, that everything the ego gets to know as being this or that (that is to say as being determined by an "essence"), is actually only an illusion of the mind deriving from the "subjective" product of the "consciousness-of." But, if it is true that the entirety of the empirical world in which we are living is in fact, as Chuang-tzu said, a Great Dream (*ta mêng*), we must, at this very moment, be deeply asleep, while thinking during our sleep that we are awake. Chuang-tzu said:

> While man is dreaming, he is not aware of the fact that he is dreaming. It is only after awakening that he understands it was really a dream.
>
> Likewise, it is only when we experience a Great Awakening that we realize that all this (that is to say all that is experienced in this world) is only a Great Dream. The stupid people, however, imagine that they are awake in fact.
>
> Actually, you and I are likewise a dream.
>
> Much more: the fact itself, on my part, of telling you that you are dreaming is itself a dream!"
>
> *Chuang-tzu*, II.

If such is the case, it is perhaps enough for man to awaken from his sleep, in order to see reality as it really is. "To awaken from sleep" is a metaphor here, for it is in fact using a new dimension of the mind while freeing the extraordinary intellectual-abstract forces, which hide and sleep in the depths of psychism and whose activity is usually hindered by the overwhelming predominance of the surface consciousness.

In the light of what I have already said about the ontological situation of empirical things, it will be easy to understand that the Great Awakening (*ta chüeh*), which is the subject here, is practically our de-objectification and de-essentialization of things. We must, in other words, create before anything else a kind of horizon of the depths in our consciousness, through which we shall be able to see things in their pre-objective way of being, in a state preceding their objectification by our "consciousness-of," that is to say before they are chained up by essential determination.

It should be observed that the kind of de-objectification of objectified things with which we are dealing here absolutely cannot take place if the "subject" remains in its state of subjectivity. The de-subjectification of the subject must also be accomplished at the same time. The annihilation of objects must be accompanied by an annihilation of the ego. The subject-object demarcation of consciousness itself must be transcended. Only after the completion of all this is the de-objectification of things accomplished, and it is then that things will be seen in their true reality.

When all things reveal themselves in their ultimate reality as totally de-objectified and de-essentialized, they are then no longer "things" in the ordinarily accepted meaning of the word. For, in such a situation, nothing proves to be such and such a thing. Consciousness stops being a "consciousness-of," since nothing remains *of* that which the mind can be conscious. The *of* being eliminated, the mind is now pure and simple "consciousness." In a perfectly significant way, Zen designates the mind having reached this stage as being a non-mind. It is a question, so to speak, of a void lighting it all up, of a void which brings everything to light, without ever objectifying: or, to put it another way, a void which brings all things to light, without there being any one "thing" to bring to light. Hung Chih Chéng Chüeh, one of the great Zen masters of the Sung dynasty, brings out in this way the extraordinary nature of the noetic activity specific to the non-mind:

> It consists in turning the *noesis* into a reality without coming in contact with things, and in bringing it all to light without objectifying it.
>
> We "know" without coming in contact with any "object" whatsoever—and this kind of *noesis* is infinitely subtle. We throw a light on everything without objectifying. Naturally, this kind of enlightenment is mysteriously profound.
>
> Being in fact an act of cognition which does not resort to the discriminating consciousness, it remains, in its oneness, beyond the subject-object demarcation. And, being a state of enlightenment, which excludes all traces of anything at all, it simply seizes everything, without seizing any one "thing," whatever it may be.
>
> *Tso Ch'an Chén*, "Words of Admonition on Meditation."

Seen through the eyes of the non-mind, all things stop being "things." The original state of the reality of being, such as this reality reveals itself to the non-mind, is characterized by ontological non-differentiation, presentation in an amorphous form, transparence, and dynamic fluidity. For nothing, from this point of view, is subject to discrimination, in relation to all the rest; nothing stands out through clearly marked "essential" limits which could surround it; nothing remains in a state of immobility, rigidly fixed in itself.

This state of things is what Chuang-tzu calls *hun tun*, or Chaos, a confusion by the non-differentiation of all things. According to Chuang-tzu, the reality of Being, *Tao*, has in itself absolutely no inner limitations. The human mind, that is "consciousness," has, however, arbitrarily "crumbled away" this very original undivided chaos into an infinity of ontological segments—certain ones big, certain ones small—while attributing to each one of them a particular name. Once the name is attributed, the respective segment takes on an illusory appearance of "essential" fixity. From that, the inarticulated Tao loses its original metaphysical purity and converts into an ontological field, made of different articulated units, a complex of ontological atoms, commonly known as being "things."

It is in the very nature of language, observes Chuang-tzu, to diversify the *Tao*, by producing everywhere—thanks to its magical power of semantic conjuring up—"essences" and "essential" divisions. All things, according to Chuang-tzu's views, are ontologically "equal" among themselves, as long as they are not differentiated one from the other by their fictitious "essences." It is clearly evident that we can only aspire to such a vision of reality if we go beyond language. The vision of reality that is then obtained will be the vision of a "vast unlimited space," where we shall notice that what are called "things" exist—seeing that they are without any "essence"—in a state of amorphous and dynamic fluidity, blending with each other without mutually blocking each other in order, finally to melt all together in a state of total non-differentiation, in the Without-Name, as Lao-tzu calls it.

This all implies, of course, that a deeply-seated mistrust of language marks the Eastern philosophical traditions. It is surprising to find constantly and fundamentally, that philosophers belonging to divergent schools hold the same negative attitude concerning language's function of articulating reality. This negative attitude is in most cases of a metaphysical nature. One begins by negating the reality of the empirical world, which would only be a creation of the human mind (that is the "consciousness-of") under the illusory influence of language, precisely in view of preparing the field for the ulterior positive construction: that of a particular kind of ontology, built on the foundation of a non-illusory vision of prelinguistic reality. In theory, nevertheless, as long as we maintain ourselves within the limits of the negative aspect, we must thoroughly keep to the negative attitude.

Among all the great Eastern philosophers, the Mahayanist Nâgârjuna (= 150–250 AD) is the one who perhaps pushed the principle of linguistic negativism to its extreme limit on the level of the theoretic possibility, by negating in a dialectical way the "essential reality"of all things, without any exception. I have already recalled, at the beginning of this study, the most basic postulate of his philosophy: *sûnyam sarvam*, which means "all things are empty." The key word here, evidently is *sûnya*, "empty," "vacant." As a technical term characteristic of his philosophical negativism, it means "void of essence," "without essence," which is the negation of a real existence of the "essential nature" (*svabhâva*).

Nâgârjuna considers the non-critical natural belief of man in the objective reality of the "things" of the empirical world as a "deeply-rooted ontological illness." And he attributes this ontological "illness" of the common man to the illusory influence of the "names," that is to language, particularly when it refers to the word's function of articulating reality. He reflects on this problem by making a particular use of the word *prapañca*.

The term *prapañca*, which in ordinary Sanskrit means diversity, diversification, or multiplicity, is, in the context of the Nâgârjunian philosophy, used in the sense of diversification (multiple articulation) of the primordial void (*sûnyatâ*, the inarticulated prelinguistic reality) in keeping with the meaning of the words. I would translate this term by "semantic diversification" or "semantic dispersion." Language "disperses" the ontological energy concentrated in the unity of the void (of what is not yet articulated) in all directions, by creating in this way, and everywhere, special ontological configurations called "things."

With its intrinsic function of semantic evocation, language, states Nâgârjuna, is basically illusory. The Vedantist Sankara, in the same way, seems to be talking about the Mâyâ-illusion of Brahmanism because it cuts up semantically the *sûnyatâ*-reality (which is in itself a whole without joints) into separate pieces, which it "essentializes"; and it does so while evoking the ontological appearance of differentiated things, which are, each one in turn, consolidated by their own *svabhâva* as such and such a thing. According to Nâgârjuna, all the things that are designated as "things" are only, in theory, hypostasized word-meanings. (As Lao-tzu would say: the Without-Name, the "uncut wood," divided into a constellation of names, the "ships.") Thus originates, under our very eyes, the empirical world.

The entire process could also be described by saying that language, from the semantic point of view, by nature conjures up (even extracts) "essences" where there are none and, consequently, that it fixates and stabilizes the *sûnyatâ*-reality, which flows eternally by dispersing it in all directions. By establishing everywhere rigidly fixed ontological boundaries it transforms the "open limitless field" of reality into a system of immobile units, of hypostasized word-meanings, which resemble corpses or fossilized forms. And it is thus that the mind of the common man considers the *sûnyatâ*-reality, which, it goes without saying, is no longer *sûnyatâ*.

Such is, briefly presented, the negative part of language criticism, so obviously characteristic not only of Nâgârjuna, but more generally of Eastern philosophy. However, "linguistic negativism" is in no way exhaustive in relation to the question as a whole. For, after the negative phase, language criticism reaches a constructive positive phase. That is so because the *sûnyatâ* itself (non-existence, void or nothingness) presents a positive aspect as well as a negative one. To be sure, in its negative aspect, the *sûnyatâ* is (or seems to be) the "nothing" in the literal sense. This "nothing" is, as Zen Buddhists like to say, "still less than a speck of dust." Nevertheless, in its positive aspect, the *sûnyatâ* is an ontological *plenum*. The *sûnyatâ*-reality, precisely because it is "absence" and "void" of anything whatsoever that is essentially fixed and determined, is in a position to determine itself, in total freedom, into a [total] phenomenal form. Pre-linguistic reality, which is the zero point of consciousness and of the being, proves to be at the same time the starting point for consciousness and for the being. In other words, the *sûnyatâ* is absolutely destructive and creative, and all at the same time. Or else, as it is often said by the Mahayana Buddhists in a highly paradoxical way, *sûnyatâ* is by itself *a-sûnyatâ*, the "void" and the "non-void."

Because *sûnyatâ* is *a-sûnyatâ*, are we thrown back once more towards the same old empirical world, the world with an ontological-semantic articulation? No. For in spite of a superficial identification, there is a deep and decisive difference between the empirical world, as it is seen at the stage of the non-critical naïve realism, and the one that is seen at the stage of the critical reaffirmation of the *a-sûnyatâ*, the latter able to be reached only after having experienced the *sûnyatâ*. What is observed in this last phase is the empirical world as it is revealed to the eyes of the consciousness at a deep level, having already experienced a total annihilation of all things.

Consciousness at the deep level, having thus already been activated, remains active even when it returns to the empirical world and continues to observe everything from the aim of the *sûnyatâ*. This must necessarily produce an extraordinary view of the empirical world. For, everything there is revealed as a contradictory unity of *sûnyatâ* and *a-sûnyatâ*, nothing and existence. A flower, for example, *is not* a flower; and *it is* nevertheless a flower. Insofar as it is *a-sûnyatâ*, reality at this stage is clearly articulated. As we have said, the ontological articulation of reality involves the semantic activity of language. But the important thing in this phase is that language be activated in such a way that it succeeds in articulating things semantically, in the empirical dimension of reality, without essentializing them. It is precisely with a *totally free*, nonessentializing articulation of reality that the function of language is manifested at the deep level, in accordance with the *a-sûnyatâ* aspect of the *sûnyatâ*.

All things are, consequently, at this stage devoid of essence. All things *are* articulated, but outside of the essence. All empirical things are *sûnyatâ* here, directly articulated in phenomenal forms (*a-sûnyatâ*). Nothing takes place between these two dimensions of reality; there is thus no possibility (no opportunity) for the "essences" to slip in. And each of the things in the empirical world is, at each instant, an *incarnation* of the ontological energy in the process of spreading itself over the entire field of the *sûnyatâ*. It is the nonarticulated in the process of being *totally* articulated, in the form of each of the empirical things. It is, as Lao-tzu would say, the Without-Name having become directly and immediately Names.

It thus seems that each of the empirical things *is* the entireness of the incarnated *sûnyatâ*-reality. A flower is a flower (*A* is *A*). But it is also all the other things (*A* is *B,C,D. . .X*). *A*, in this sense, is non-*A*. Zen expresses this idea more poetically: "In a speck of dust the entire universe is contained," and "A flower blooms and the entire world blossoms in the form of spring."

For the surface consciousness ("consciousness-of"), the empirical world is an ontological structure whose constituent units are definitely disjointed. It is a world dominated by the principle of mutual obstruction among all things. For the deep, nonobjectifying, nonessentializing consciousness, the empirical world appears to be, on the contrary, completely transformed. There is no longer any ontological obstruction, for all things lacking essences are *free*. Infinitely transparent, they are open to each other. Seen through the eyes of the "consciousness-of," things were ontologically opaque and closed to each other. Now, their essential boundaries having been removed, they are found in a state of ontological luminosity and transparence. Being translucid without any essence, they freely melt into each other, so much so that the entire universe seems to be an iridescent fabric of interpenetrating lights.

All "things," precisely in this sense, are ontologically ambivalent. Every "thing" is reduced to nothingness, but by being "nothing," it *is* a "thing."

Contributors

Jean E. Charon, Theoretical Physicist, University of Paris; Director, Center of Research on Complex Relativity and Relations to Consciousness, Villebon-sur-Yvette, France.

Diane Cousineau, Psychologist, Research in Creativity, University of Quebec, Canada.

José M.R. Delgado, Biologist, Director, Department of Research, Ramon y Cajal Center, Madrid, Spain.

Willis W. Harman, Sociologist, Senior Social Scientist, SRI International; President, Institute of Noetic Sciences, Menlo Park, California.

Toshihiko Izutsu, Professor of Oriental Philosophy, University of Keio, Toyko.

Jean Lerède, Charge de Cours, University of Montreal; University of Sherbrook, Montreal, Canada.

Edgar Morin, Sociologist, Director of Research, National Center for Scientific Research, Paris, France.

Henryk Skolimowski, Professor of Philosophy, University of Michigan, Ann Arbor, Michigan.

Jerzy A. Wojciechowski, Professor of Philosophy, University of Ottawa, Ottawa, Ontario, Canada.

Also at the presentation of papers and participating in the discussions were Mitsuo Ishikawa, René Berger, Elizabeth Rauscher, Ashok Gangadean, Michel Maffesoli, and Francoise Bonardel.

Sources

The content of this volume was derived from papers presented in Committee VI ("Spirit and Science: Imaginary and Reality") at ICUS XIII ("Absolute Values and the New Cultural Revolution"). The symposium, one of seven sponsored by ICUS, was held at the J.W. Marriott Hotel in Washington, D.C., September 2–5, 1984, and was chaired by Jean E. Charon.

Index

This index is of proper names and major themes in the book. A number in brackets behind a page number refers to a footnote on that page. The page numbers in **bold face** type refer to pages written by the person indexed.

J

K

L

M